Geometric Tolerancing

A Text-Workbook

Geometric Tolerancing

Second Edition

Richard S. Marrelli

Late Professor of Industrial Technology

Los Angeles Pierce College
Woodlawn Hills, California

Patrick J. McCuistion, Ph.D.

Department of Industrial Technology
Ohio University
Athens, Ohio

GLENCOE

McGraw-Hill

New York, New York   Columbus, Ohio   Mission Hills, California   Peoria, Illinois

Library of Congress Cataloging-in-Publication Data

Marrelli, Richard S.
       Geometric tolerancing: a text-workbook / Richard S. Marrelli,
    Patrick J. McCuistion.—2nd ed.
            p. cm.
       Includes index.
       ISBN 0-02-801882-6 (text)
       1. Tolerance (Engineering) I. McCuistion, Patrick J. II. Title.
TS172.M37 1997
620′ .0045—dc20

                                                                       95-41232
                                                                          CIP

Geometric Tolerancing: A Text–Workbook, Second Edition

Printed in the United States of America.

Send all inquiries to:
Glencoe/McGraw-Hill
936 Eastwind Dr.
Westerville, OH 43081

ISBN 0-02-801882-6

2 3 4 5 6 7 8 9 10 11   045   03 02 01 00 99 98 97

CONTENTS _____

PREFACE

INTRODUCTION

Geometric Tolerancing Second Edition is intended to fill the needs of students just beginning the study of geometric dimensioning and tolerancing and many other users by not being too complex nor too simplistic. This text has been updated to reflect the major changes in the latest revision of the ASME Y14.5M-1994 Dimensioning and Tolerancing standard. Previous users of this text will notice that the two datum chapters (4 and 20) have been combined to form a single datum Chapter 4. A chapter on symmetry has been added. Many of the problems in the back of the text have been modified for better readability or are completely new.

ORGANIZATION

This text is arranged by geometric characteristics: the simpler characteristics are covered first, followed by the more complex. The illustrations are not meant to be complete but to show only the necessary information for the concept at hand.

The problems and exercises at the back of the book are thought provoking and intended to spark more in-depth discussions about the topics. They are designed to be used as study guides.

TEACHER'S MANUAL

A teacher's manual provides teaching guidelines for how this text can be used. There are answers to all problems and exercises, periodic tests, and a final exam.

ACKNOWLEDGMENTS

The author wishes to express his thanks to John Beck, Marianne L'Abbate, Bruce Albrecht, Jan Hall, and the graphics staff from Glencoe. He would also like to thank his wife Dolores for putting up with his late-night schedule.

If you have suggestions for improvement of this text, please send them to the author in care of Glencoe, 936 Eastwind Drive, Westerville, Ohio 43081.

Patrick J. McCuistion
Athens, Ohio

About the Author

Prior to joining Ohio University in 1989 in the Department of Industrial Technology, Dr. McCuistion worked for many years in various engineering design, drafting, and checking positions at several manufacturing industries. He also taught for three years at Texas A&M University. He has provided instruction in geometric dimensioning and tolerancing to many industrial, military, and educational institutions. He has also written articles and given presentations at national meetings on the topic.

Dr. McCuistion is an active member of several ANSI subcommittees, including Y14.5 Dimensioning and Tolerancing, Y14.3 Multiview and Section View Drawings, Y14.35 Revisions, Y14.36 Surface Texture, and B89.3.6 Functional Gages. He is also active in the Society of Manufacturing Engineers, the National Association of Industrial Technology, and the American Society for Engineering Education.

CHAPTER 1

Introduction to Geometric Tolerancing

This text is intended for students of drafting, design engineering, manufacturing, and quality control, as well as for working professionals in these areas. It is assumed that the reader already has an understanding of the concept of dimensional tolerance in engineering drawings. This text is meant to expand the concept of tolerance (total permissible error) to include the *geometry* of a part as well as its dimensions.

1-1 WHAT IS A GEOMETRIC TOLERANCE?

The word *geometry* in relation to an object refers to two characteristics:

1. The shape (e.g., cylindrical, rectangular, triangular, and so forth).
2. The relationship of its features. A feature is any portion of an object, such as a surface, hole, or groove.

For example, a pulley may have a cylindrical hole with a keyway. This describes the shape of the opening. The specified relationship may require that the keyway be centered on the axis of the hole and that the hole be perpendicular to one face of the hub. The opposite hub face may also need to be parallel with the first face. These examples of geometric characteristics must be specified on the drawing and interpreted correctly by manufacturing and quality control personnel.

The example in Fig. 1-1 will be used to illustrate why it is necessary to control the geometry of a part as well as its dimensions. The function of this part is to space two other parts .750 apart, within ±.01. Since all the tolerances on the spacer in Fig. 1-1 are ±.01, the part may be .01 smaller or larger in width and height. These tolerance zones are shown with phantom lines in Fig. 1-2. The zones are shown larger than actual size for clarity. The limits of the part can be anywhere within the phantom lines. The example shown in Fig. 1-3, although exaggerated, is possible given the dimensions.

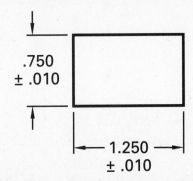

Fig. 1-1 Spacer.

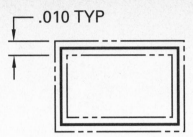

Fig. 1-2   Tolerance zones.

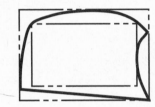

Fig. 1-3   Possible part geometry.

The part produced as shown is acceptable as far as the dimensional tolerances are concerned. But the variation in the shape and relationship may mean the part will not function properly. Geometry can be controlled by specifying tolerances on the form of the surfaces and the relationship between surfaces. This may be done using local notes as shown in Fig. 1-4. Figure 1-5 illustrates how symbols can be used to show the same information presented in Fig. 1-4. Now, regardless of the dimensional variation, the top and bottom surfaces must be flat within .001 and parallel within .002.

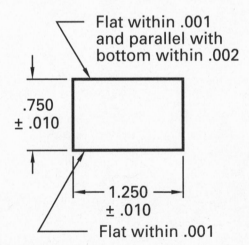

Fig. 1-4   Geometric control using local notes.

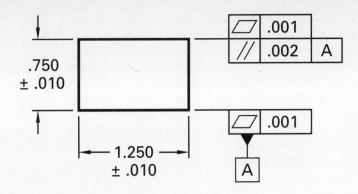

Fig. 1-5   Geometric control using symbols.

Geometric tolerance, then, is the permissible error in the geometric characteristics of an object—its form and the relationship of its features.

1-2   A BRIEF HISTORY

Prior to World War II, geometric dimensioning was generally not specified on engineering drawings. Any responsibility for geometric form or relationship was left to the people who made the parts. When an increased level of precision was required, this procedure was inadequate. It caused many rejects and/or much time devoted to reworking parts so they could be assembled. Gradually designers began to specify the required geometric dimensions using local notes.

The use of the symbol system known today started with the British just before World War II. Because of the collaboration among British, Canadian, and American firms during the war, the method became known to North American engineers and was gradually adopted by North American industry. In 1966 the United States of America Standards Institute (USASI), now the American National Standards Institute (ANSI), published a standard procedure for the application of geometric form and position tolerances: ANSI Y14.5-1966.

A *standard* is a document specifying certain methods and procedures to be used by all interested parties in a certain industry. For example, the Society of Automotive Engineers (SAE) has standards on the sizes and materials of rubber hose for automotive vehicles, and these standards are used by all the manufacturers of automobiles, trucks, and tractors.

1-3   ASME Y14.5M-1994

Prior to 1966, different industries maintained their own dimensioning standards, which caused problems for vendors who supplied those industries. They were forced to create drawings using different standards for each. It also created problems in the development of international standards.

The International Organization for Standardization (ISO) was created in 1946. It was difficult for the United States to speak with one voice at ISO meetings when it did not have one single standard. This situation changed in 1966 with the release of the Y14.5 standard. The United States has now become one of the innovators in international standards. Although there are still some differences between U.S. and ISO dimensioning and tolerancing standards, each group has greatly influenced the other.

The Y14.5-1966 standard was revised in 1973 and again in 1982. The "M" designation was added in 1982 to indicate that all the dimensional values are metric. The latest revision of the standard, finalized in January 1994 and approved for publication in December 1994, is known as ASME Y14.5M-1994. The American Society for Mechanical

Engineering (ASME) is the group responsible for publishing the standard. Even though the Department of Defense (DoD) adopted this standard in March 1994, they still maintain their version of the standard in DoD-STD-100, Section 10M.

One of the difficulties sometimes encountered by users of drawings is trying to determine which standard was invoked when the drawing was done. This information should be noted on the drawing but sometimes is not. With the escalating use of the Y14.5 standards through the years, and the differences from revision to revision, it may not be enough to know just the current standard. Appendix A provides a review of some of the principal differences between earlier versions of the Y14.5 standard. This text is in conformance with the ASME Y14.5M-1994 standard.

CHAPTER 2

Symbols

2-1   GEOMETRIC CHARACTERISTIC SYMBOLS

ASME Y14.5M-1994 standardizes fourteen geometric characteristic symbols, in addition to three modifier symbols and several other symbols to define exact geometric requirements on drawings. The symbols are shown in Fig. 2-1. All the geometric characteristic symbols should be memorized, which should be easy to do because the shape of each symbol reminds one of the characteristic. The symbol for parallelism, for example, is two parallel lines. Perpendicularity is symbolized by two perpendicular lines. Two concentric circles are used for concentricity. The others are equally simple.

The ASME Y14.2M-1992 Line Conventions and Lettering standard states that free-hand dimensions should be .125 inches (3.5 mm) tall for drawings 17 × 22 inches in size and smaller, and .156 inches (5 mm) tall for drawings larger than 17 × 22 inches. The proportional sizes relative to letter height "h" are shown in Fig. 2-1. Notice that all the angles are 30°, 45°, or 60°. The symbols may be drawn easily with triangles and a circle template, or with a special template made for this purpose. Templates are available with symbol sizes based on lettering heights from $\frac{1}{8}$ (3 mm) to $\frac{5}{32}$ (4 mm) to $\frac{3}{16}$ (5 mm).

The use of the symbols and their application to drawings is explained in Chapters 7 through 19. Figure 2-2 presents a quick résumé of some of the data in these chapters.

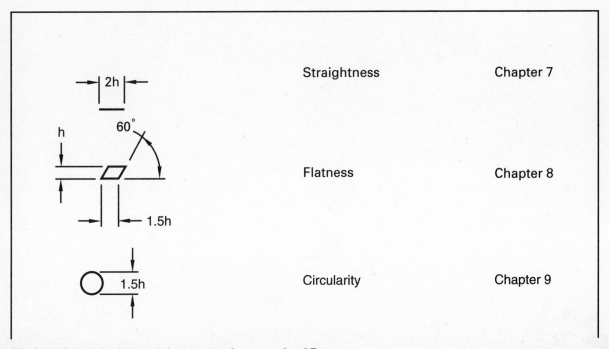

Straightness		Chapter 7
Flatness		Chapter 8
Circularity		Chapter 9

Fig. 2-1   Geometric characteristics (*continued on pages 6 and 7*).

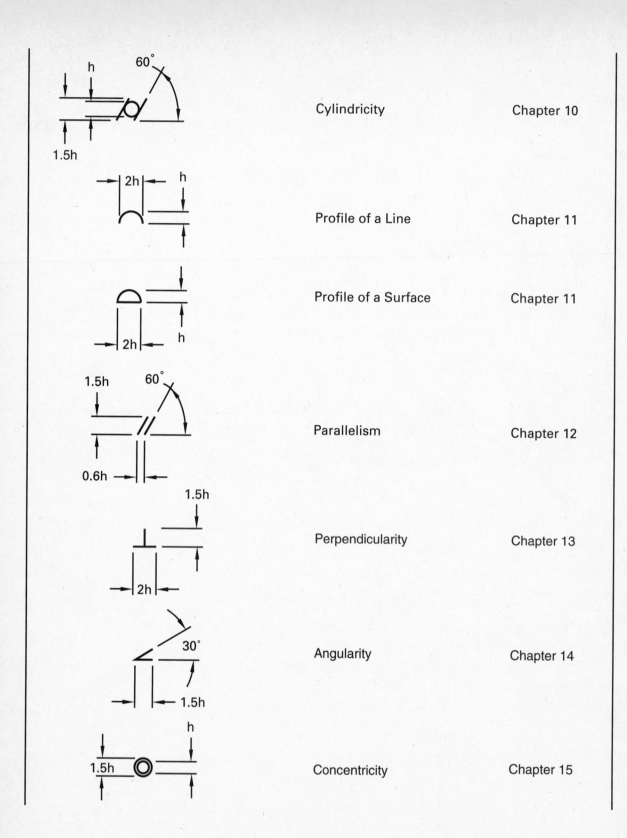

Cylindricity   Chapter 10

Profile of a Line   Chapter 11

Profile of a Surface   Chapter 11

Parallelism   Chapter 12

Perpendicularity   Chapter 13

Angularity   Chapter 14

Concentricity   Chapter 15

Fig. 2-1 Geometric characteristics (*continued*).

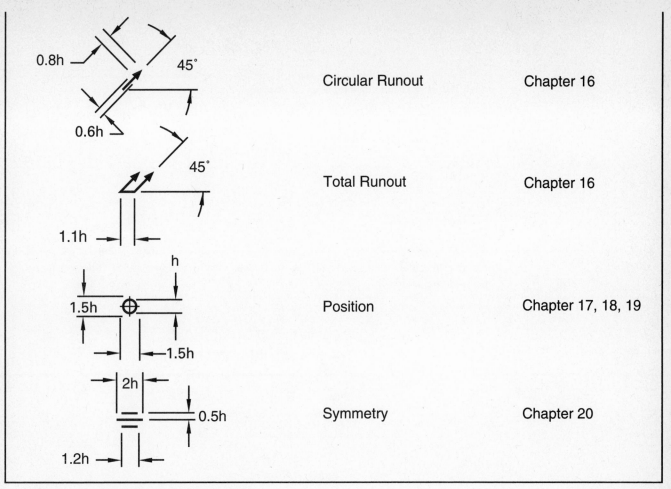

		Circular Runout		Chapter 16	
		Total Runout		Chapter 16	
		Position		Chapter 17, 18, 19	
		Symmetry		Chapter 20	

Fig. 2-1 Geometric characteristics (*continued*).

Geometric Characteristic	Symbol	Type of Tolerance	Pertains to:	Use of Modifier with Feature Tolerance	Use of Modifier with Datum
Straightness	—	Form	Individual feature only	Modifiers not applicable	No datum
Flatness	⟋⟋				
Roundness	○				
Cylindricity	⌀				
Profile of a Line	⌒	Profile	Individual feature or related features		If a size feature, RFS is implied, unless Ⓜ or Ⓛ is specified.
Profile of a Surface	⌓				
Angularity	∠	Orientation	Related features	If a size feature, RFS is implied, unless Ⓜ or Ⓛ is specified.	
Perpendicularity	⊥				
Parallelism	//				
Circular Runout	↗	Runout		RFS is implied. If design requires that feature be Ⓜ, use position tolerance (⊕).	RFS is implied. If design requires that datum be Ⓜ, use position tolerance (⊕).
Total Runout	↗↗				
Concentricity	◎	Location		RFS is implied unless Ⓜ or Ⓛ is specified.	RFS is implied unless Ⓜ or Ⓛ is specified.
Symmetry	≡				
Position	⊕				

Fig. 2-2 Résumé of data on geometric characteristics.

2-2   FEATURE CONTROL FRAME

A complete feature control frame is shown in Fig. 2-3. It is made up of a box (frame) divided into compartments containing, at the least, a geometric characteristic symbol and a geometric tolerance value. The frame is read left to right and may contain the following:

1. Geometric characteristic symbol
2. Tolerance
3. Tolerance modifier
4. Datum
5. Datum modifier

When required, a modifier symbol is given after the tolerance. When a datum is specified, the appropriate letter is given in the next compartment, and this may be followed by its own modifier symbol.

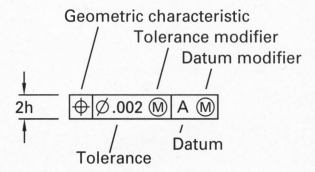

Fig. 2-3   Complete feature control frame shown actual size.

The feature control frame is drawn with thin lines. The feature control frame is always drawn horizontally on the sheet, like local notes. It is applied only once to a specific feature of a part.

A feature may have more than one geometric tolerance. For example, flatness and parallelism may be applied to one plane surface. Figure 2-4 shows different examples of feature control frames. The applications of the frames are shown in the chapters dealing with the specific geometric characteristics. The feature control frame is directed to the feature being controlled by one of the methods shown in Fig. 2-5.

The symbol is read:

▱ .002

ENGLISH

The flatness of the feature must be within .002 in.

○ 0.05

METRIC

The circularity of the feature must be within 0.05 mm.

Fig. 2-4   Examples of feature control frames. *(continues)*

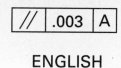

ENGLISH

The parallelism of the feature must be within .003 in. relative to datum feature A.

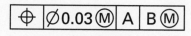

METRIC

The position of the feature axis must be within 0.03 at MMC relative to datum feature A and datum feature B at MMC.

Fig. 2-4 Examples of feature control frames. *(continued)*

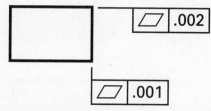

(a) Attach a side of the frame to a horizontal or vertical extention line drawn from an edge view of the feature.

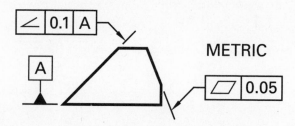

(b) Attach a corner of the frame to a slanted extension line.

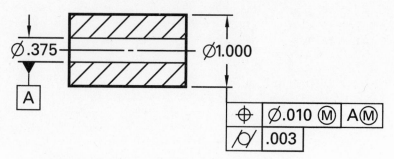

(c) Attach a side of the frame to a dimension line. This is used only when the feature is a *size* feature (a cylinder, keyway, tab, etc.).

Fig. 2-5 Methods of applying feature control frames and datum identifying symbols to drawings. *(continues)*

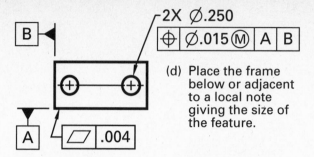

(d) Place the frame below or adjacent to a local note giving the size of the feature.

(e) Attach the left or right side of the frame at its midheight to a leader that points to the controlled feature or to an extension line drawn from it.

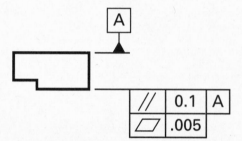

(f) Stack or hang the frame on another frame pertaining to the same feature. Draw the frames so that the right end of a shorter frame aligns with a compartment line of a longer frame.

Fig. 2-5   Methods of applying feature control frames and datum identifying symbols to drawings. *(continued)*

CHAPTER 3

Terms

Some of the terms used in geometric tolerancing are defined in this chapter. A definition may be somewhat specialized for this subject and different from general English use. The terms are given in the approximate order in which they occur in the text.

3-1   FEATURE

A *feature* is any portion of an object. A feature may be a point, an edge, a center line, or a plane or curved surface. It can also be a size, such as the width of a slot, groove, or tab, or the diameter of a cylinder, in which case it is called a *size feature*.

3-2   ERROR

An *error* is an unintentional variation from a desired dimension or geometric form, location, or orientation. The error is acceptable and is not wrong unless it exceeds limits (tolerances) given on the drawing.

3-3   ELEMENT

An *element* is any line, real or imaginary, that can be drawn on a surface, including flat surfaces (planes), curved surfaces (cylinders, cones, spheres) and irregular surfaces. All the elements of a plane are straight lines in any direction. Elements of a cylinder may be circles, all with the same diameter as the cylinder (circular elements), or they may be straight lines parallel to the axis (longitudinal elements). Elements of a cone may be circles, all of different diameters, on imaginary planes perpendicular to the axis, or they may be straight slanted lines on the surface, converging at the apex. Elements of irregular surfaces may be lines parallel to (and on either side of) a line representing the theoretically exact, true profile of the irregular surface.

3-4   RADIAL LINE

The word *radial* has the same root meaning as the words *ray* and *radiate*. It means pointing directly toward or away from a center. The sun's rays are radial lines, for example, because they extend outward from a center, the sun. A leader on a drawing is a radial line when it is directed to a circle because it points to the center. A radial line does not have to intersect the center; it is just the direction of it that makes it radial.

3-5   DATUM

A *datum* is a theoretically exact feature from which dimensions may be taken or from which the geometric form or position of another feature may be determined. The use of datums in connection with geometric tolerances is explained in Chapters 4 and 20.

3-6   TOLERANCE ZONE

A *tolerance zone* is the area of permissible error defined by a dimension. The shape of a tolerance zone depends on the feature and how it is controlled. It may be two-dimensional (the area between two parallel lines) or three-dimensional (the space between two parallel planes or a cylinder). Other shapes may also be defined, but these are not common.

3-7   BASIC DIMENSION

A *basic dimension* may have two definitions. When it relates to general toleranced dimensions, it is a nominal dimension from which upper and lower limits are derived. When it relates to geometric dimensioning, it is a theoretically exact dimension that locates a tolerance zone. A basic dimension does not have a tolerance. It is enclosed in a rectangle that is two times the letter height and is wide enough to contain the dimension. Figure 3-1 shows some examples of geometric basic dimensions.

Fig. 3-1   Geometric basic dimensions.

3-8   MAXIMUM MATERIAL CONDITION (MMC)

Maximum material condition (MMC) is the condition in which a size feature (a hole, slot, or groove, or a shaft, pin, boss, or tab) contains the most material (see Fig. 3-2). For a shaft or other solid feature, it will be the *maximum* permitted diameter, the size in which the most material is contained. For a hole or other hollow feature, the MMC is the *minimum* permitted diameter, wherein the hole is the smallest and the object, again, contains the most material.

When maximum material condition is used as a modifier in a feature control frame, the tolerance specified applies only when the controlled feature or datum is at its maximum material condition (size). This concept will be explained in more depth in Chapters 6 through 19.

3-9   LEAST MATERIAL CONDITION (LMC)

Least material condition (LMC) is the situation in a size feature when the feature contains the least material—the smallest shaft; the largest hole (see Fig. 3-2). When least material condition is used as a modifier in a feature control frame, the tolerance specified applies only when the controlled feature or datum is at its least material condition (size). Applications of the use of LMC are given in Chapter 17, Section 17.6.

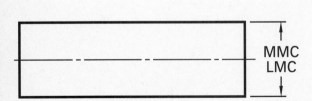

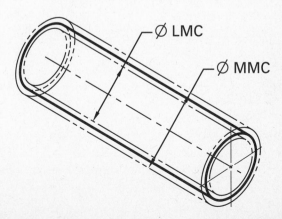

A size dimension controls the size and the form of a part. The size and form may vary within an envelope between the least and maximum material conditions.

Fig. 3-2   Least and maximum material conditions.

3-10   REGARDLESS OF FEATURE SIZE (RFS)

Regardless of feature size is the default condition for all geometric tolerances and any size datums. It means the tolerance is applied no matter what the size of the feature. This default condition will be explained in more detail in Chapters 6 through 19.

3-11   VIRTUAL CONDITION

The *virtual condition* (VC) is the collective effects of the size and any geometric tolerance of form, orientation, or position. The following are calculations for the virtual conditions of a shaft and a hole:

Shaft: VC = maximum diameter + geometric tolerance

Hole: VC = minimum diameter − geometric tolerance

Any size feature that has a geometric dimension applied will have a virtual condition.

3-12   FULL INDICATOR MOVEMENT (FIM)

The *full indicator movement* (FIM) is the total movement (reading) of the pointer of a dial indicator when used to test a geometric tolerance. The dial indicator is an inspection device commonly employed to measure geometric errors. A detailed explanation of the dial indicator is given in Chapter 5. The terms *full indicator reading* (FIR) and *total indicator reading* (TIR) were formerly used for, and had identical meaning to, FIM.

3-13   PROJECTED TOLERANCE ZONE

A *projected tolerance zone* is a position tolerance zone that starts at the surface of a part and extends outward. Most position tolerance zones start at the surface of the part and extend inward. The length of the zone is specified either as a dimension away from the surface of the part with a chain line or in the feature control frame. In either instance, the projected tolerance zone symbol is placed in the feature control frame after the geometric tolerance. An example is shown in Fig. 3-3. Applications are shown in Chapter 18.

Fig. 3-3   Projected tolerance zone applied to a feature control frame.

CHAPTER 4

Datums

4-1 DEFINITIONS

A *datum* is a theoretically exact point, line, axis, or area from which geometric measurements are taken. A theoretical datum is established by the contact of a datum feature and a simulated datum (see Fig. 4-1). The *datum feature* is any physical portion of a part. The *simulated datum* is what the datum feature contacts and should imitate the mating part in the assembly. A simulated datum may be a mounting surface of a machine tool, a surface of an assembly fixture, or a surface of an inspection holding fixture. Neither the datum feature nor the simulated datum are perfect, but the simulated datum should always be more precise than the datum feature because the datum is generated by the contact of the datum feature and the datum simulator.

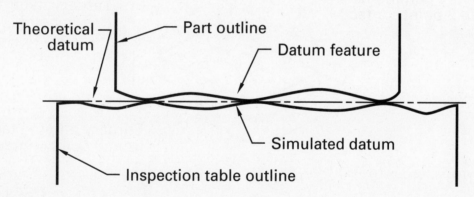

Fig. 4-1 Datum creation.

4-2 DATUM REFERENCE FRAME

An unsupported object in space can move in an infinite number of ways. If a part is to be machined or inspected, it must be restricted from moving. The different movement possibilities are called *degrees of freedom*. The degrees of freedom can be divided into six linear and six rotational directions (see Fig. 4-2). Contact between the datum features and simulated datums restricts these degrees of freedom. The motions are listed below:

	Linear Motions		**Rotation Motions**	
X direction	1—left	2—right	7—positive	8—negative
Y direction	3—front	4—rear	9—positive	10—negative
Z direction	5—up	6—down	11—positive	12—negative

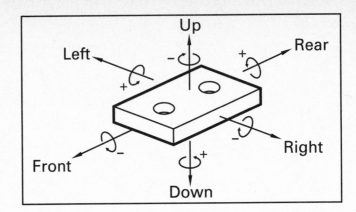

Fig. 4-2   Twelve degrees of freedom.

The *datum reference frame* is the restrictive environment in which a part is placed. The datum reference frame should allow parts to be made, measured, and used with consistency. For rectangular parts, the datum reference frame may consist of three mutually perpendicular planes, as shown in Fig. 4-3. For circular parts, the datum reference frame may also consist of three planes (see Fig. 4-4). Contact with the plane surface creates the primary datum. Contact with the circular feature establishes two planes perpendicular to themselves and the primary plane. The intersection of these planes creates the axis of the part, which is a theoretical datum.

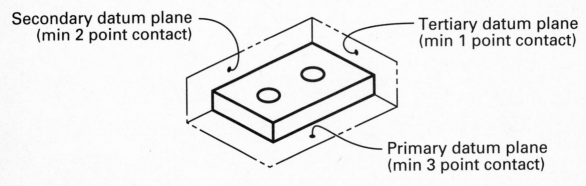

Fig. 4-3   Datum planes—rectangular parts.

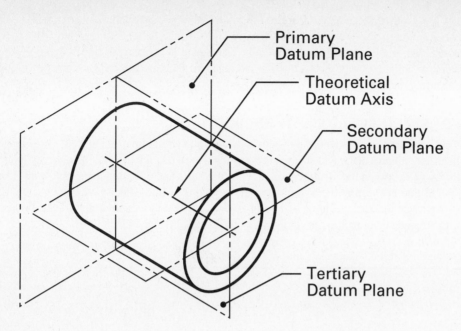

Fig. 4-4   Datum planes—circular parts.

4-3   SELECTION OF DATUM FEATURES

When selecting features to serve as datum features, observe the following rules:

1. Select functional features. An example would be a cylindrical surface that supports the part in a bearing.
2. Select corresponding features on mating parts (features that contact each other).
3. Select features that are readily accessible for manufacture and inspection.

While it is desirable, it is not always possible to select finished surfaces as datum features. In many cases specific points, lines, or areas of a surface are identified as datum targets. Datum targets are specified where datum features are rough, uneven, or on different levels, such as on castings, forgings, or weldments.

4-4   PLANE DATUMS AND SIZE DATUMS

Datum features may be of two different types: plane features or size features. Plane features have two-dimensional area, but no three-dimensional size. A flat surface is a plane feature. Most contoured surfaces are considered plane features.

A size feature has either an external or internal diameter (pin or hole) or an external or internal width (tab or slot). The size of these datum features must be considered when you are geometrically designing parts. The geometric control may be applied when the datum feature is as large as it can be, as small as it can be, or regardless of its size. These conditions are called maximum material condition (MMC), least material condition (LMC), and regardless of feature size (RFS), respectively. The table below presents the different conditions for internal and external features.

Feature	MMC	RFS	LMC
Internal	Smallest size	Largest actual mating size	Largest size
External	Largest size	Smallest actual mating size	Smallest size

To determine which condition to consider is one of the cornerstones of geometric dimensioning and tolerancing (GDT) datum application. If there is a clearance between the parts, consider MMC. If there is an interference between the parts, consider RFS. If there is a concern for a minimum wall thickness, consider LMC. More information will be provided about these conditions in later chapters.

External rectangular size features are often positioned using a gripping device like a vise, especially if it is referenced at RFS. A vise can also be used to position an internal rectangular size feature. External circular size features can be positioned with a lathe chuck or collate, or a V-block. Internal circular size features can be positioned using an expanding mandrel.

4-5   SPECIFYING DATUMS ON DRAWINGS

Datum features can be identified using the symbol shown in Fig. 4-5. Figure 4-6 illustrates the different ways plane and size features can be identified on a drawing. Notice the subtle differences between the identification of a plane feature and a rectangular size feature. The straight line connecting the triangle and the square is in line with, or attached to, the dimension line for a size feature but not for a plane feature. For a circular feature, the bottom of the triangle that touches the circle is drawn as a straight line tangent to the circle.

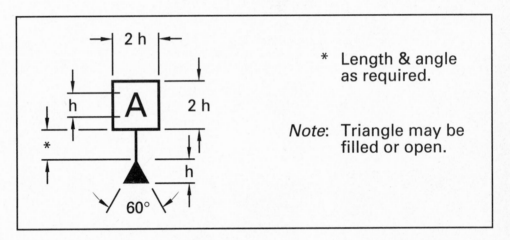

Fig. 4-5   Datum feature symbol. h = letter height.

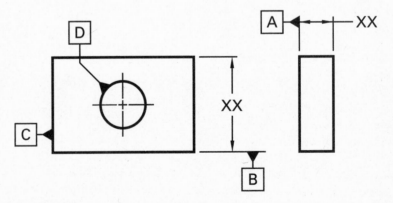

Fig. 4-6   Datum feature symbol use.

The letters in the square boxes do not have to be in alphabetical order. The important part of the letter is how it is used in the feature control frame.

4-6   SPECIFYING DATUM TARGETS ON DRAWINGS

If specific portions of a feature will be used to establish the theoretical datums, they should be identified with datum target symbols. The size and different parts of the symbol are shown in Fig. 4-7.

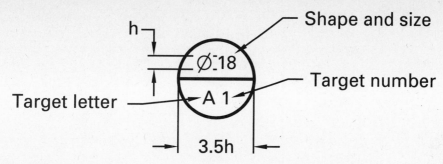

Fig. 4-7   Datum target symbol.

There are three items of interest in the examples that follow. First, datum targets can be located with either basic or toleranced dimensions. If basic dimensions are used to specify the locations, the actual tolerances for those dimensions are derived by someone other than the designer or drafter. If toleranced dimensions are used, the designer or drafter retains control of the target locations. Second, the leader from the datum target symbol does not use an arrow to point to the part. Third, if the actual target is on a surface hidden from view, the leader line is drawn as a dashed line.

Figure 4-8 illustrates how a datum target point is specified on a drawing. The cross is two times the letter height with a 90° included angle. It is common practice to contact the part with a round-headed pin.

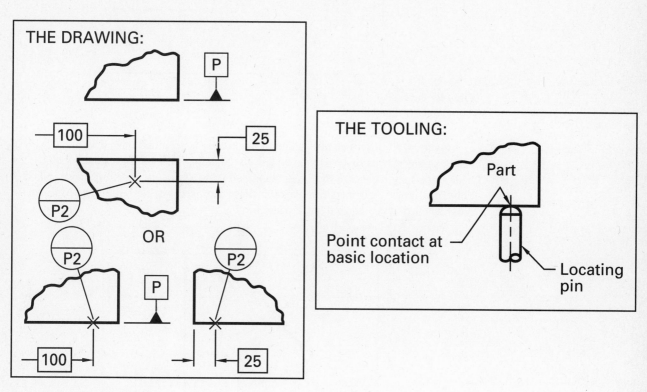

Fig. 4-8   Datum target point.

Figure 4-9 illustrates how a datum target line is specified on a drawing. The line is drawn as a phantom line. In a view where the line would appear as a point, it is shown with the 90° cross. The contact with the part is achieved by the side of a pin.

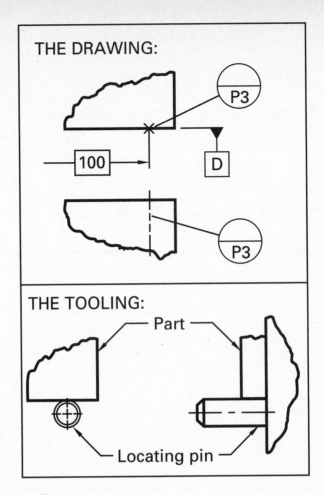

Fig. 4-9   Datum target line.

Figure 4-10 illustrates how a datum target area is specified on a drawing. This area can be of any size or shape as long as it is less than the full size of the datum feature. The area shape may be shown on the drawing as in part (a) or its position can be indicated only by a cross, as in part (b). Phantom lines are used for the outline, and the area is cross-hatched with a general hatch pattern.

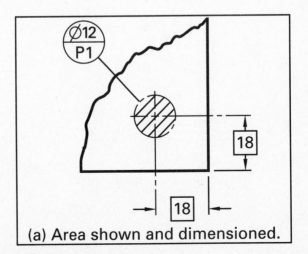

Fig. 4-10   Datum target area. *(continues)*

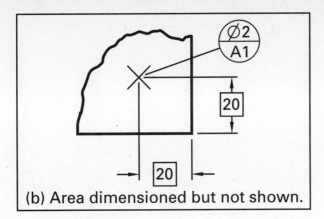

(b) Area dimensioned but not shown.

Fig. 4-10 Datum target area. *(continued)*

4-7 DATUM USE EXAMPLES

Figures 4-11 through 4-15 show examples of how the different datums can be used to position parts for production, inspection, and assembly. The example shown in Fig. 4-11 is a fully machined cover plate. The datums will be established by the full contact of the datum features (surfaces A, B, and C) with simulated datums of equal or greater size.

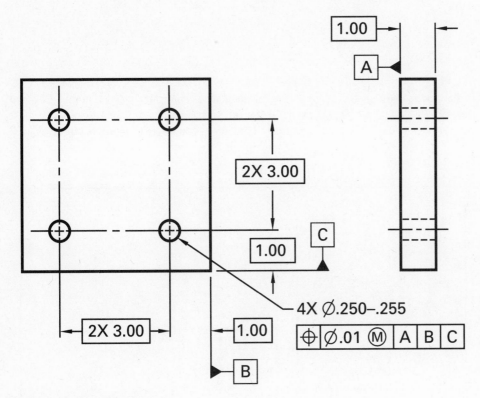

Fig. 4-11 Machined cover plate.

The part shown in Fig. 4-12 is a cast cover plate with only one surface machined (datum feature A). Datum feature A is specified as a full surface because it is a mating surface. Datum features B and C are identified with datum target points. They are not functional surfaces but are used to provide complete three-dimensional restriction. The combination of full surfaces, datums, and datum targets to position parts is common for castings and forgings.

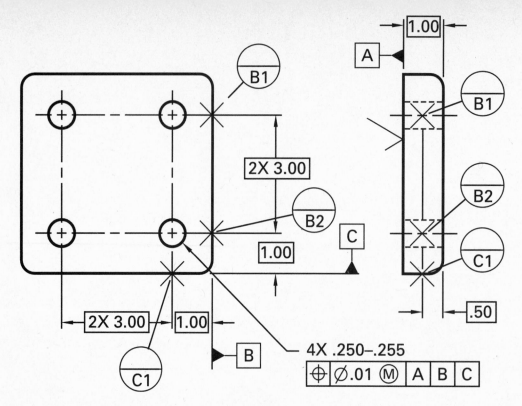

Fig. 4-12   Cast cover plate.

Figure 4-13 illustrates an example of an aligning pin. The .495–.500 diameter is controlled to the datum A face. The .245–.250 diameter is controlled to the datum A face and the .495–.500 diameter. In this example, the full surfaces of datum features A and B must be used to create the datums. The maximum material condition of datum feature B is used to center the part.

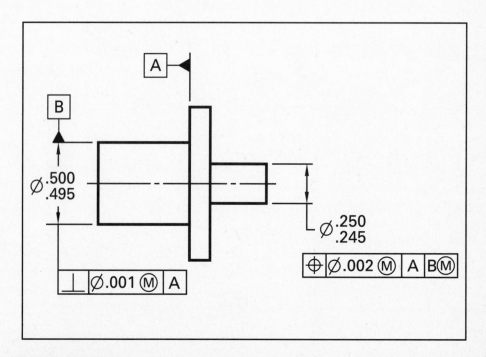

Fig. 4-13   Aligning pin.

The support hub shown in Fig. 4-14 uses three datum target points (B1, B2, and B3) to center the part. These points must be contacted in a regardless-of-feature-size condition to create datum B. Datum feature A is used only to establish datum feature C as the first machined surface. Datum feature A may be considered manufacturing information—it is used only to manufacture the part, and it is not necessary information for the part's function.

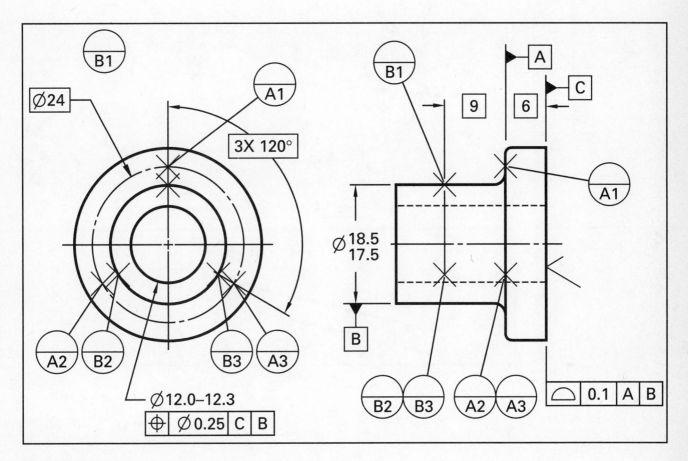

Fig. 4-14   Support hub.

The connecting link shown in Fig. 4-15 illustrates the use of step datums and equalizing datums. A step datum consists of datum targets placed on two or more different surfaces that restrict the same degrees of freedom. Datum A is comprised of two points on one surface and one point 20 mm away. The 20 mm distance can be a toleranced or basic dimension or it can be geometric controlled.

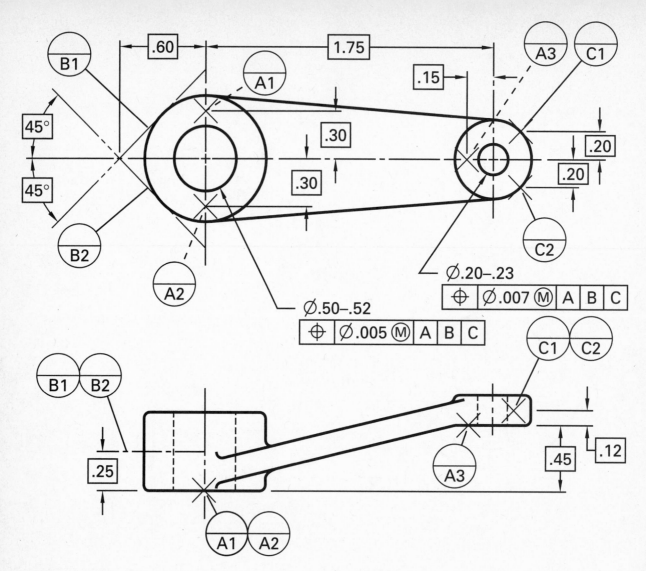

Fig. 4-15   Connecting link.

Equalizing datums consist of specific datum targets that, when combined, center non-circular parts. Datum targets B1, B2, C1, and C2 in Fig. 4-15 establish a center plane through the middle of the part. This example illustrates the potential difficulty of trying to identify three mutually perpendicular planes for the datum reference frame. The main question to answer is, "Have all the necessary degrees of freedom been restricted so that the part will function properly?" If the answer is yes, the appearance or how the datums have been identified doesn't matter.

CHAPTER 5

Inspection of Geometric Tolerances

5-1 INTRODUCTION

Inspecting size tolerances involves measuring side-to-side distances (differential measurements). It also includes verifying that the form of the part remains within the boundary described by the maximum material condition (see Chapter 6). Geometric tolerances are different. They are either contained within size dimensions or related to different features of the part. For this reason, geometric tolerances require different inspection methods and equipment. This chapter will describe two common measuring devices and discuss how they are used.

5-2 THE DIAL INDICATOR

The *dial indicator* is a device used to detect movements as small as one ten-thousandth of an inch or one thousandth of a millimeter. It consists of a glass-covered dial about the size of a small pocket watch, and a pointer that revolves over the dial. Attached to the pointer with a mechanical link is a spring-loaded stem or probe that extends from the side or back of the case. Small movements of the probe are registered by the pointer and can be read on the dial.

A typical dial indicator is illustrated in Fig. 5-1. The pointer of this indicator will rotate 180° for a movement of the probe of .012 inch up or down. The complete movement of the probe is about .100 inch in several revolutions of the pointer. The bezel (outside rim) is attached to the dial and can be rotated by hand to align zero on the dial with the pointer wherever it happens to be when the probe is pressed against the surface to be tested. Zero does not have to be at the 12 o'clock position.

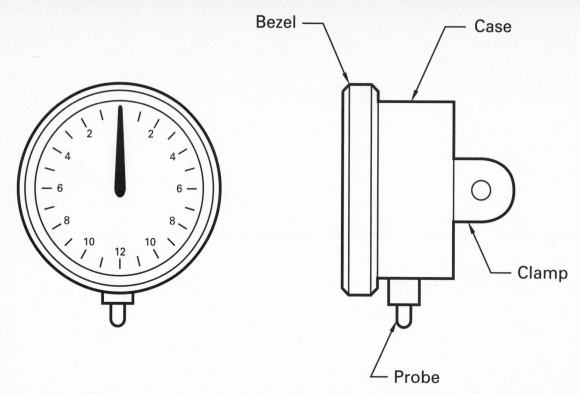

Fig. 5-1   A typical dial indicator.

5-3   HOW THE DIAL INDICATOR IS USED

In a typical application, for example, testing parallelism, the part to be tested is placed on a surface plate and the dial is clamped to a support that is free to slide on the surface plate alongside the part (see Fig. 5-2). The dial indicator is positioned above one end of the surface to be tested and lowered on its support until the probe is in contact with the surface. The probe will move upward a small amount, causing rotation of the pointer. Now the bezel is rotated by hand to "zero" the dial. To test for parallelism, the dial indicator is moved across the test surface by sliding the base of the support over the surface plate. If the upper surface is not perfectly parallel with the lower surface resting on the surface plate, the probe will move up or down from its starting point. This movement will be registered by the rotation of the pointer. The total movement of the pointer in both directions (full indicator movement, or FIM) is the parallelism error of the two surfaces. In the setup shown in Fig. 5-2, the full indicator movement is .003, which is equal to the parallelism error.

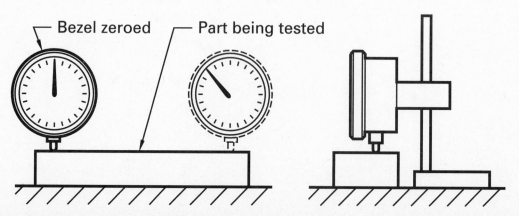

Fig. 5-2   A surface plate setup for testing parallelism.

5-4 THE COORDINATE MEASURING MACHINE

A coordinate measuring machine (CMM) is also used to detect small amounts of variation (see Fig. 5-3). The instrument that detects the location of a surface is a small ruby ball probe that is mounted to a vertical beam (larger CMMs usually have a horizontal beam). This beam can move up and down (Z axis). The vertical beam is mounted to a horizontal beam that can move from side to side (Y axis). The horizontal beam is supported by two vertical colums that move in and out (X axis) and rest on a granite table (the tables on larger CMMs are usually steel). This table is called a *surface plate.* The X, Y, and Z dimensions of the probe are noted on a digital display or computer screen. The CMM is a precise and expensive inspection tool.

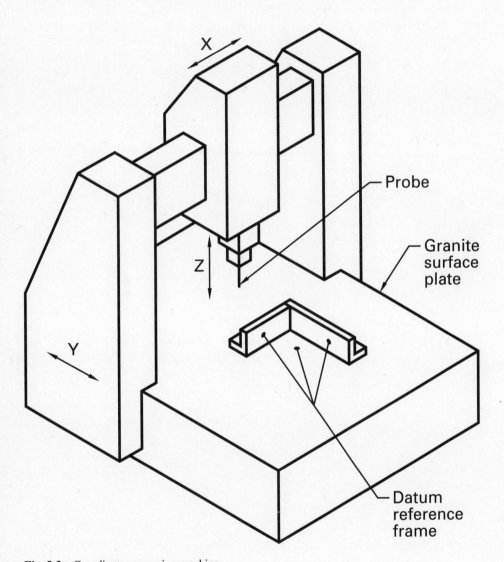

Fig. 5-3 Coordinate measuring machine.

5-5 HOW THE COORDINATE MEASURING MACHINE IS USED

The dial indicator is small and can be mounted easily on a machine tool where the part is produced. The CMM is stationary, and parts are brought to it for inspection. To measure parallelism with the CMM, it must be run first through a calibration program so that the location of the center of the ruby probe is known. Next, the top of the surface plate must be determined by touching the ruby probe on the surface in at least three different locations. Each of the probe locations can be stored in a computer as X, Y, and Z dimensions. All the probe locations are used to create a geometric plane. The part is placed on the surface plate. The opposing surface of the part is probed with the ruby ball in at least three locations. This derived plane is compared to the plane developed from the probe locations of the top of the surface plate. Any discrepancies are noted and printed for reporting purposes.

5-6   WHEN TO USE INSPECTION DEVICES

The dial indicator or CMM can be used for measuring variation in most of the geometric characteristics. Several figures in the remaining chapters show the dial indicator being used to inspect flat, curved, and irregular surfaces. The CMM could easily be substituted. It takes time to learn how to use each of these devices properly. When manufacturing large quantities of parts, different inspection methods can be used. Many factors must be considered when determining which measuring method and device to use. This text will focus on the different geometric controls and will use inspection illustrations for explanation purposes only.

CHAPTER 6

General Rules

6-1 INTRODUCTION

There are four general rules that apply automatically when dimensioning parts, unless exceptions are specified in the feature control frame or in a note. The rules are only noted in this chapter; they will be discussed in more detail in later chapters.

6-2 LIMITS OF SIZE RULE

There are three parts to the *limits of size rule*. First, a feature of size must be within the stated upper and lower size limits. At any cross-section, the part must not exceed the maximum material condition (MMC) or least material condition (LMC). Second, the feature may not exceed a perfect form boundary at MMC. If the part were produced at the MMC, it cannot have any deviation in form. Third, the part form may vary between the LMC and the MMC.

The limits of size rule actually refers to the "envelope principle," which states that there is an envelope of size between the LMC and MMC within which all the surface elements must lay. The envelope principle is used in the United States and allows for the calculation of an allowance and maximum clearance without the addition of any geometric tolerances. Figure 6-1 illustrates the envelope principle with two mating parts—a shaft into a hole. Even if the shaft and hole are tapered or bowed to the extreme limit of the MMC, they will still fit together because they have not exceeded the MMC.

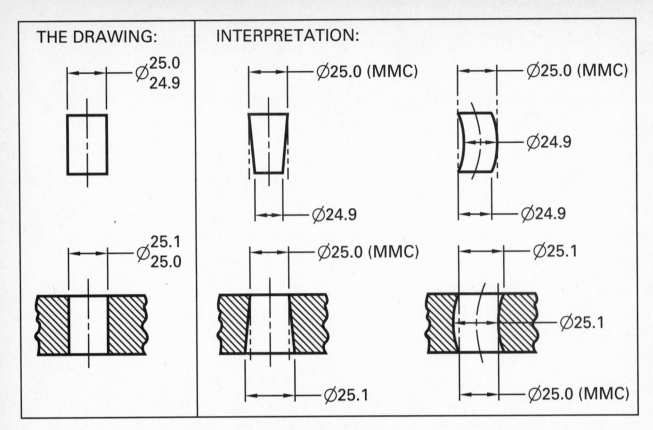

Fig. 6-1   How the limits of size rule applies to an external and an internal feature.

The limits of size rule does not apply to extruded or rolled shapes such as bar stock, sheet stock, tubing, or structural beams. The industries that make these stock parts have their own established standards for size and form. This rule also does not apply to the relationship between size or plane features. Geometric dimensions must be applied to specify feature-to-feature relationships.

6-3   REGARDLESS OF FEATURE SIZE RULE

The *regardless of feature size* (RFS) *rule* states that all geometric tolerances and any size datums are applied RFS unless otherwise stated. If the tolerance or datum is applied at MMC or LMC, it must be noted with the appropriate symbol. MMC or LMC modifiers cannot be applied to flatness, circularity, and cylindricity, circular runout, total runout, concentricity, or symmetry.

6-4   PITCH DIAMETER RULE

The *pitch diameter rule* states that all geometric tolerances and any size datums are applied to the axis derived from the pitch cylinder of threaded features, unless otherwise specified. If the major or minor diameter of the thread is the datum feature, then it must be noted on the drawing (e.g., MINOR DIA or MAJOR DIA). The placement of this note is most often under the feature control frame or datum feature symbol.

For any feature other than threads, the specific datum feature must be specified. This applies to parts like gears and splines. The note placement is similar to that for threads.

6-5   VIRTUAL CONDITION RULE

The *virtual condition* size of a part is the combination of the MMC and the geometric tolerance controlling that feature. A size feature that has previously been geometrically controlled and is used as a datum applies at its virtual condition, even though MMC is specified. This concept is used when there is a clearance between the datum feature and its mating part (see Fig. 6-2).

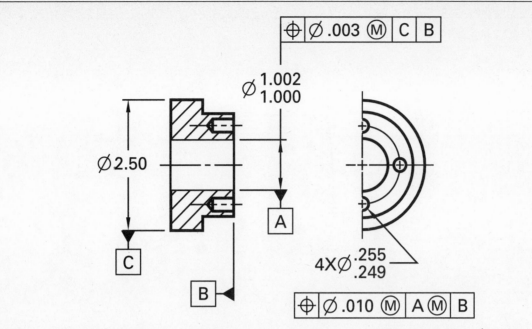

Datum A, a size feature, applies at its virtual condition: MMC ($\varnothing$1.000) minus positional tolerance ($\varnothing$.003) equals $\varnothing$.997. When A is produced larger than its MMC, the position of the $\varnothing$.255–.249 holes, as a group, is allowed to shift by the difference between the actual size and the MMC size.

Fig. 6-2   Virtual condition.

Datum feature B in Fig. 6-2 must be considered at its virtual condition when you are inspecting the four holes. The perpendicularity for datum B relative to datum A means the hole may be tilted by .003. If the mating part is controlled in the same manner, the largest mating shaft that will fit in the 1.000–1.002 diameter hole is actually .997. Not only is the .997 diameter the size of the largest mating shaft, it is also the size of the pin in a functional gage used to inspect the part.

CHAPTER 7

Straightness

7-1 DEFINITION

Straightness error is the measure of how much each element in a surface or the axis of an object deviates from being a straight line. Straightness can be applied to a single surface, such as a plane or cylindrical surface, or it can be applied to a size feature, such as the diameter of a cylinder, in which case the effect is quite different, as shall be shown.

7-2 ELEMENT STRAIGHTNESS—PLANE SURFACES

The tolerance zone is the area between two parallel straight lines. The tolerance zone must be contained within the size limits of the dimension that includes the surface. Straightness is considered a refinement tolerance. The straightness tolerance must be less than the size tolerance.

The straightness tolerance applies to *all* the elements of the surface in the direction indicated by the placement of the feature control frame in the appropriate view. The straightness tolerance applies only in the direction shown in that view. See Fig. 7-1. A straightness tolerance can be applied in two directions by placing feature control frames in two adjacent views, as shown in Fig. 7-2.

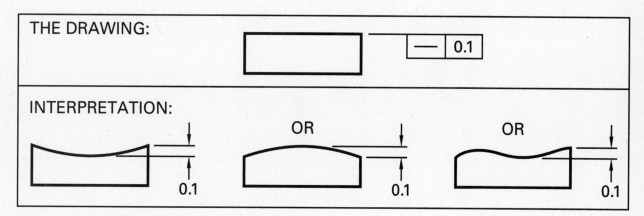

Fig. 7-1   Application of element straightness to a plane surface.

Fig. 7-2   Application of straightness in two directions.

7-3 ELEMENT STRAIGHTNESS—CYLINDRICAL SURFACES

To control the longitudinal elements on the surface of a cylinder, the feature control symbol is applied to the straight line representing those elements or to its extension line, as in Fig. 7-3, but not to the diameter or its dimension line. A straightness tolerance thus applied controls all the longitudinal elements of the surface of the cylinder but does not control the axis. The tolerance zone is the same as for plane surfaces. It is the area between two parallel straight lines.

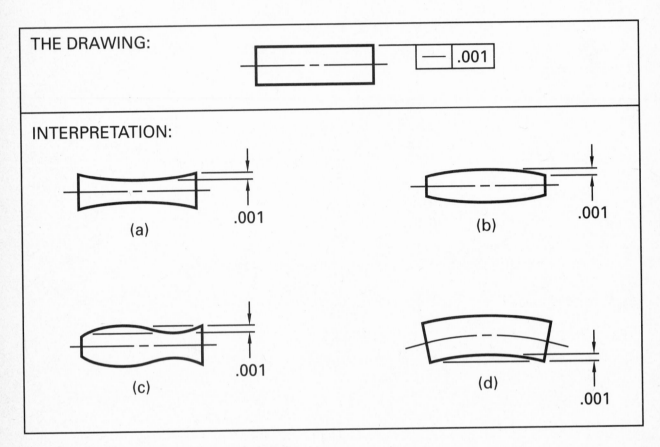

Fig. 7-3 Application of element straightness to a cylinder.

The out-of-straightness of an object may occur as shown in Fig. 7-3 (d), where all the elements are bowed in the same direction. Obviously, the axis must also be bowed, but the control is not on the axis—only on the surface elements.

The surfaces of other shapes that are symmetrical about their axes, such as cones, square and hexagonal bars, and so forth, are treated the same way as cylindrical surfaces.

7-4 STRAIGHTNESS OF SIZE FEATURES

It is sometimes necessary to control the straightness of a whole size feature rather than just its surface elements. This is done on the drawing by applying the straightness symbol to the size dimension or its dimension line. Two examples are shown in Fig. 7-4.

THE DRAWING:

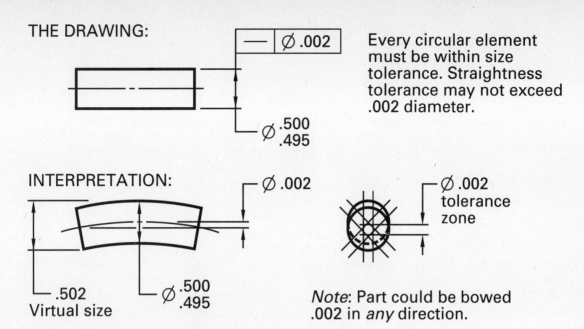

Every circular element must be within size tolerance. Straightness tolerance may not exceed .002 diameter.

Note: Part could be bowed .002 in any direction.

(a) No modifier specified; RFS implied.

THE DRAWING:

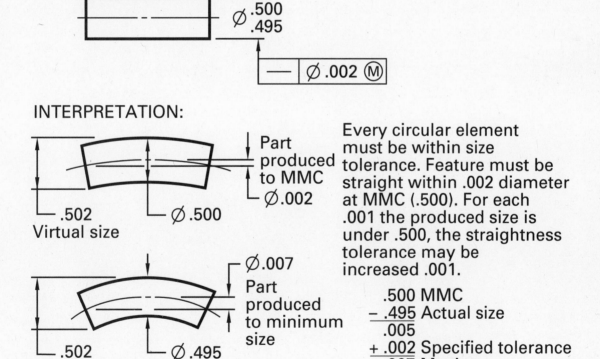

Every circular element must be within size tolerance. Feature must be straight within .002 diameter at MMC (.500). For each .001 the produced size is under .500, the straightness tolerance may be increased .001.

.500 MMC
– .495 Actual size
.005
+ .002 Specified tolerance
.007 Maximum straightness tolerance

(b) MMC specified.

Fig. 7-4   Straightness applied to a size feature.

The object shown in part (a) of Fig. 7-4 might be a long drive shaft that fits into bearings only at the ends. The purpose of the straightness tolerance is to limit out-of-balance due to bowing of the shaft. When a shaft or pin is to fit in a hole for its entire length, MMC may be specified as shown in part (b). This ensures that the part will fit, while allowing additional straightness tolerance if the part is not produced at its MMC.

When straightness is applied to a size feature, the form of the feature may exceed the perfect form boundary at maximum material condition. Straightness is the only geometric control that will allow this form violation of the limits of size rule.

7-5   UNIT STRAIGHTNESS

In some instances, it is desirable to specify straightness on a unit basis. For example, the straightness tolerance can be specified for one inch of length or three inches, or 50 mm or 100 mm, and so forth. An example might be an especially long, slender rod. The tolerance might be specified as .0005/1.00 (which means that the tolerance is .0005 for each one inch of length) or .01/25 (.01 mm for each 25 mm). (See Fig. 7-5.) The reason for specifying straightness on a unit basis is to prevent the possibility of all the straightness error occurring in one area. Straightness of cold-drawn shafting from the mill, for example, is specified in thousandths of an inch per foot.

When specifying unit straightness, take care that the *total* straightness error is not excessive. A straightness tolerance of .001/1.00 results in .004 in two inches, .009 in three inches, and so forth. A 20-inch feature on this basis could be .400 out of straightness because the chord height of an arc is proportional to the *square* of the chord, which is illustrated in Fig. 7-5 (a). It is often necessary, therefore, to specify a *total* straightness in addition to the straightness per specified length, as shown in Fig. 7-5 (b).

THE DRAWING:

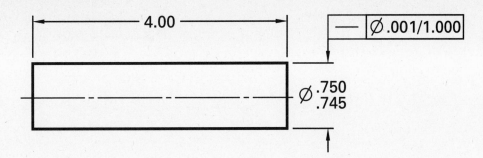

INTERPRETATION:

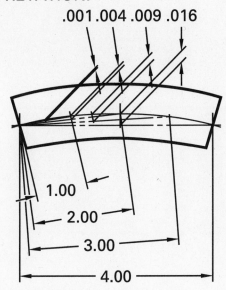

.001 .004 .009 .016

The size feature is to be straight within .001 for each 1.000 inch of length. The total straightness error could be as much as .016.

(a) Unit straightness only.

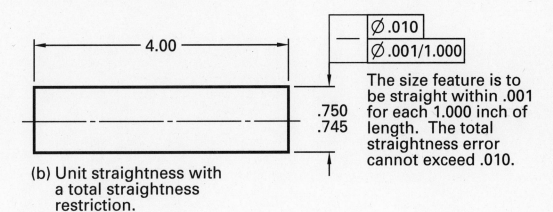

The size feature is to be straight within .001 for each 1.000 inch of length. The total straightness error cannot exceed .010.

(b) Unit straightness with a total straightness restriction.

Fig. 7-5 Straightness specified on a unit basis.

CHAPTER 8

Flatness

8-1 DEFINITION

Flatness error is the measure of how much a plane surface deviates from being a true plane. Whereas straightness affects the elements of a surface in one direction, flatness affects all the elements in all directions. A surface can be straight in one direction but not be flat. Figure 8–1 is an example. Notice that all the elements in the long direction (1, 2, 3, 4) are straight, but the elements in the short direction (a, b, c, d) are all curved. The surface is not flat.

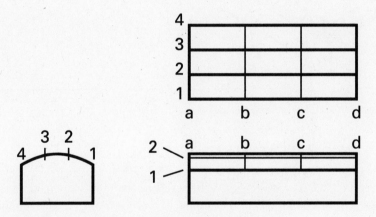

Fig. 8-1 A surface may be straight in one direction and not be flat.

A flatness tolerance zone is the space between two parallel planes (see Fig. 8-2). The tolerance zone must be contained within the size limits of the dimension that includes the surface. Flatness is considered a refinement tolerance. The flatness tolerance must be less than the size tolerance.

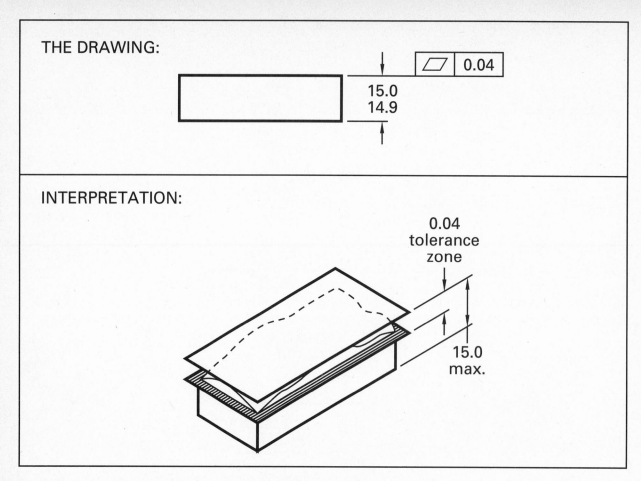

Fig. 8-2 Specification of flatness and the shape of the tolerance zone (metric).

8-2   SPECIFYING FLATNESS ON DRAWINGS

Flatness is specified by directing the feature control frame to the edge view of the surface, as shown in Fig. 8-2. The entire surface being controlled must be within the size tolerance, and no element of the surface may extend beyond the MMC boundary of the part.

8-3   UNIT FLATNESS

It is sometimes desirable to specify flatness on a unit basis. Because flatness is an area control, the unit is specified as a width and depth, e.g., 1.00 × 1.00. (See Fig. 8-3.) This specification prevents the possibility of all the machining flatness errors occurring in one area.

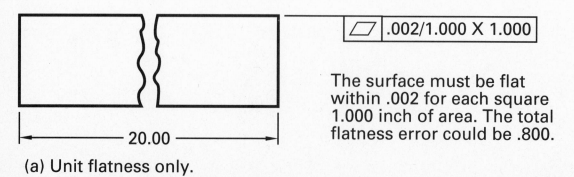

The surface must be flat within .002 for each square 1.000 inch of area. The total flatness error could be .800.

(a) Unit flatness only.

Fig. 8-3 Flatness specified on a unit basis. *(continues)*

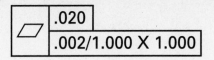

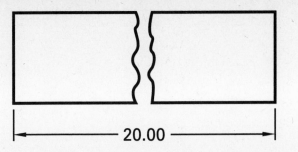

The surface must be flat within .002 for each square 1.000 inch of area. The total flatness error cannot exceed .020.

(b) Unit flatness with a total flatness restriction.

Fig. 8-3   Flatness specified on a unit basis. *(continued)*

8-4   FLATNESS IN A LOCALIZED AREA

If the flatness occurs only in a localized area, the area is outlined with a chain line on the surface. The area is then cross-hatched and dimensioned. A leader is attached to the flatness callout and the other end terminates inside the area with a dot (see Fig. 8-4). This application is most often used on castings or forgings for localized machining.

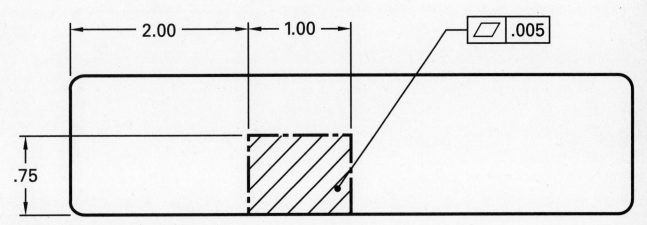

Fig. 8-4   Flatness applied to a specified area.

CHAPTER 9

Circularity

9-1 DEFINITION

A *circle* is a curved line in which every point is equidistant from a point. This point is called the center. *Circularity error* is the measure of how much a circle deviates from having all its points equidistant from the center. It is measured as the radial area between two perfect concentric circles within which every point on the circle being tested must lie. The circularity for an exaggerated imperfect circle is shown in Fig. 9-1. Note that the center of the two perfect circles is not necessarily the same as the theoretical center of the circle, the center shown on the print. In inspecting for circularity, neither center is ever actually located.

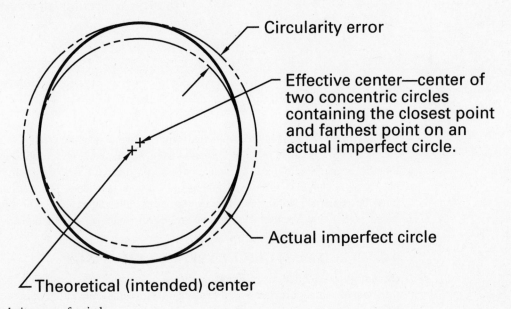

Circularity error

Effective center—center of two concentric circles containing the closest point and farthest point on an actual imperfect circle.

Actual imperfect circle

Theoretical (intended) center

Fig. 9-1   Circularity error of a circle.

Circularity error is measured in a plane perpendicular to the axis of a circular feature, which may be a cylinder, a cone, or some other surface that is circular in cross-section but may vary in diameter (see Fig. 9-2). In specifying circularity on any circular surface, the circularity error of every circular element in the surface is controlled at every cross-section. Each circularity inspection is independent of all other circularity inspections.

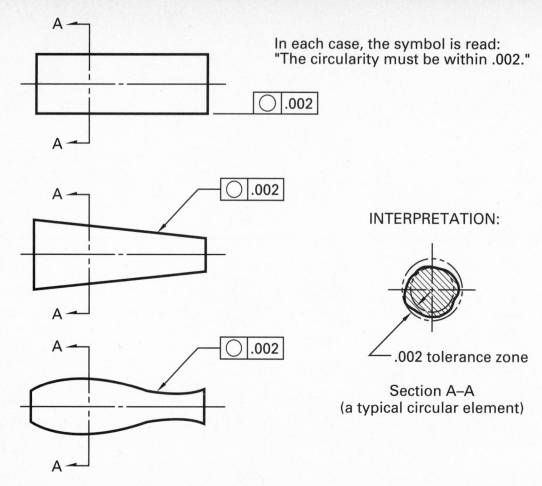

Fig. 9-2 Application of circularity tolerance to circular surfaces.

The tolerance zone is the radial area between two concentric circles. It is not a diameter or a radius because it is not measured from the center. A circularity tolerance can be thought of as a straightness tolerance curled into a circle. Whereas straightness affects only straight elements, circularity affects only circular elements. The circularity tolerance must be contained within the dimensional size limits; therefore, it is referred to as a *refinement of size.* The tolerance will always be less than the size tolerance.

9-2   SPECIFYING CIRCULARITY ON DRAWINGS

Circularity is specified on drawings as shown in Fig. 9-2. For a cylinder, the feature control symbol can be directed with a leader to the circular view rather than to the rectangular view when that is more convenient—but the symbol is never directed to both the circular view and the rectangular view.

Special circularity inspection equipment is available to test the part. Typically, the part is centered on a turntable. A dial indicator or an electronic sensor is placed in contact with the surface and the table is rotated (see Fig. 9-3). The position of the part on the table is adjusted until the smallest readings are obtained from the indicator—when the effective axis of the part is aligned with the axis of the turntable. The turntable is then rotated to test the circularity of the part. The indicator readings every 30° or so are plotted on a large-scale circular graph until the part has been rotated 360°. The graph thus obtained is a profile of the particular circular element tested (Fig. 9-4). A transparent film overlay of concentric circles of a known distance apart at the same enlarged scale is placed over the graph. The circularity error is the distance between any two concentric circles that contain between them all the points of the profile (see Fig. 9-4).

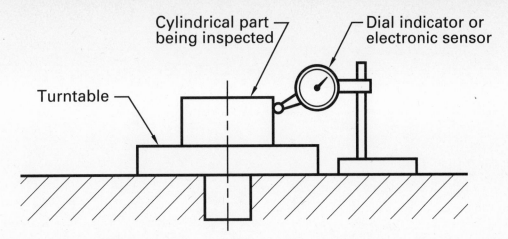

Cylindrical part being inspected

Dial indicator or electronic sensor

Turntable

Fig. 9-3   Setup for inspecting circularity.

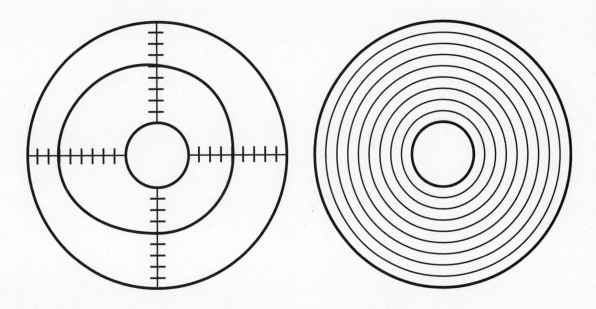

Circular graph with profile of circular element plotted: Each graduation represents .001 inch.

Transparent overlay: The radial distance between concentric circles represents .001 inch.

If the transparent overlay were placed over the graph, we would see that the circularity error of this circular element is .002.

Fig. 9-4   Circular graph and transparent overlay at the same enlarged scale.

This procedure tests the circularity of just one circular element. A sufficient number of other elements are then tested to obtain a clear idea of the circularity of all the circular elements of the surface.

Other methods of testing for circularity include revolving the part on lathe centers, on an arbor, or in a chuck, and taking readings on a dial indicator without the use of a graph. However, these are not as accurate because the dial indicator reads eccentricity between the tooling and the part and bowing of the part, in addition to circularity error. These methods are acceptable if the full indicator movement (FIM), as tested, is within the specified circularity error. But if the FIM is greater, it is not known how much of the indicator movement represents circularity error. Testing in vee-blocks is not recommended because the angle of the vee and the imperfect profile of the feature can combine to reduce or increase the apparent circularity error.

CHAPTER 10

Cylindricity

10-1   DEFINITION

Cylindricity is a combination of circularity and straightness. A perfect cylinder has every circular element perfectly round and every straight element perfectly parallel to the axis. *Cylindricity error* is any deviation in an actual part from this perfection.

10-2   CYLINDRICITY TOLERANCE ZONE

The tolerance zone is the radial space between two concentric perfect cylinders within which every straight element and every circular element must lie (see Fig. 10-1). A cylindricity tolerance can be thought of as a flatness tolerance curled into the shape of a cylinder. Flatness controls all the straight elements of a plane surface in both directions, and cylindricity controls all the straight elements of a cylindrical surface in the direction of the axis and all the circular elements in a perpendicular direction. The cylindricity tolerance must be contained within the dimensional size limits; therefore, it is referred to as the *refinement of size*. The tolerance will always be less than the size tolerance.

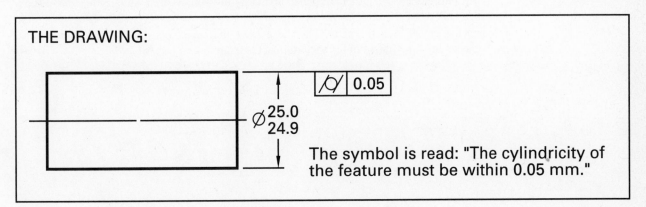

Fig. 10-1   Specifying cylindricity (metric) (*continues*).

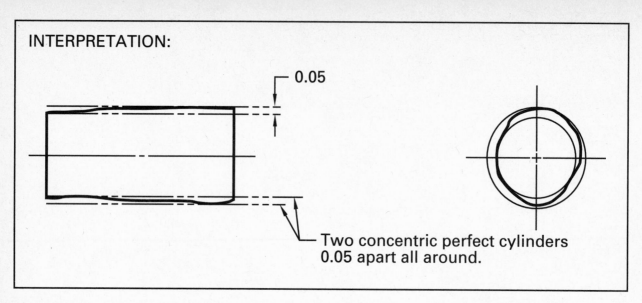

INTERPRETATION:

0.05

Two concentric perfect cylinders
0.05 apart all around.

Fig. 10-1   Specifying cylindricity (metric).

10-3   SPECIFYING CYLINDRICITY ON DRAWINGS

Cylindricity is usually specified as shown in Fig. 10-1. An alternate method is to direct a leader from the mid-height of the feature control frame to the cylindrical surface, either in the circular or the rectangular view.

10-4   MEASURING CYLINDRICITY ERROR

Cylindricity is inspected using the same type of device used to check circularity. The main difference is that the individual circularity inspections are no longer separate inspections. Each of the circular inspections are plotted on the same circular graph while the tracing device remains parallel to the axis of the part. This measuring technique also checks for straightness of the longitudinal elements.

CHAPTER 11

Profile

11-1 DEFINITION

A *profile* is the outline or contour of an object. The surface shown in profile can be flat, curved, or irregularly curved in one direction, or it can be a compound curve. The desired profile is usually a theoretically exact contour defined with basic dimensions. The profile error is any deviation of an actual part from the desired profile.

Because the true position of the surface is defined with basic dimensions, profile controls size as well as form. Profile is the only geometric control that can control size.

There are two different types of profile: profile of a line and profile of a surface. Profile of a line is a two-dimensional control used when the elements of an irregular surface in one direction only must be controlled. Profile of a surface is a three-dimensional control that may be used to control any of the surfaces mentioned in the previous paragraph in all directions.

11-2 PROFILE TOLERANCE ZONE

For a profile of a line tolerance, the tolerance zone is the area between two imaginary curved lines perfectly parallel to the desired profile. For a profile of a surface tolerance, the tolerance zone is the space between two imaginary curved surfaces parallel to the surface of the desired profile.

The tolerance zone can be specified using one of the methods shown in Fig. 11-1. The example shows the profile of a surface symbol in the feature control frame. The zone can be applied on a bilateral or unilateral basis. Unless otherwise specified, it is assumed to be bilateral, half on the inside and half on the outside.

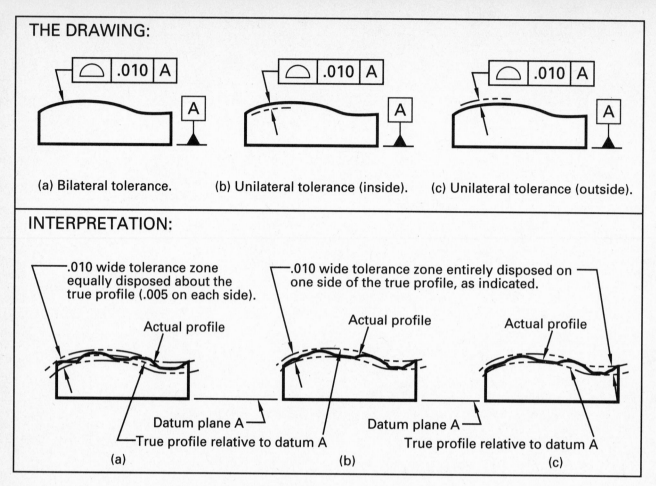

Fig. 11-1 Specifying profile tolerance—surface tolerance shown.

11-3   SPECIFYING PROFILE TOLERANCES

To specify a bilateral tolerance, a feature control frame is drawn with a leader pointing to the edge view of the surface. When a unilateral tolerance is required, a line representing the boundary of the tolerance zone is drawn inside or outside the basic profile line, as applicable. This line is drawn as a phantom line a short distance away from the profile line. A dimension line indicating the width of the tolerance zone is drawn and a feature control frame is attached to this dimension line. See Fig. 11-1, parts (a) and (b).

Usually the profile specification is shown in a view where the surface appears as an edge. The desired profile is defined by basic dimensions, as shown in Figs. 11-2 and 11-3.

THE DRAWING:

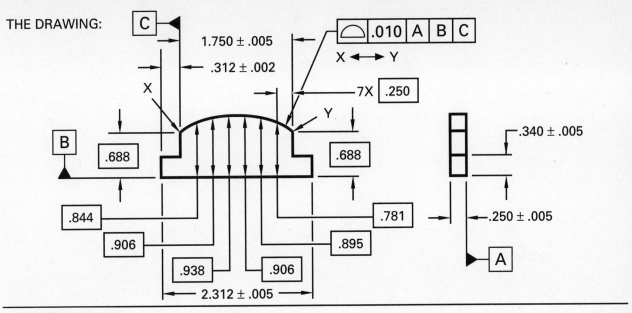

INTERPRETATION:

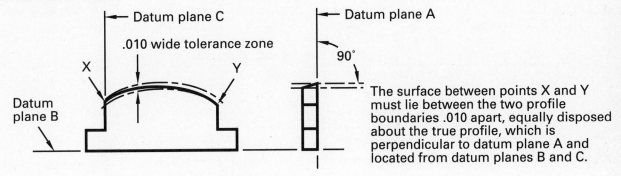

Datum plane C

.010 wide tolerance zone

X   Y

Datum
plane B

Datum plane A

90°

The surface between points X and Y must lie between the two profile boundaries .010 apart, equally disposed about the true profile, which is perpendicular to datum plane A and located from datum planes B and C.

Fig. 11-2 Specifying profile of a surface between points.

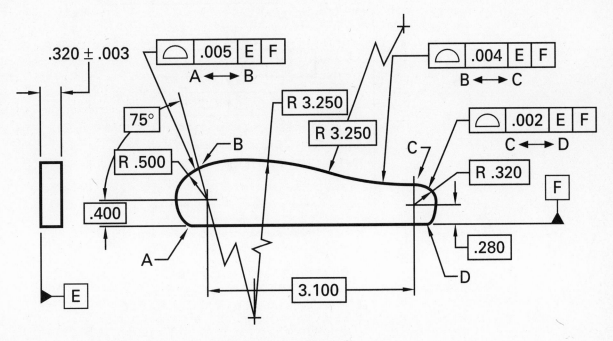

Fig. 11-3 Specifying different tolerances on different portions of the profile.

The profile tolerance usually applies only to a specific portion of the surface. Letters are placed at the beginning and ending points of the portion of the surface to be controlled. These points are noted below the feature control frame by placing a line with arrows on either side pointing to the letters, as shown in Figs. 11-2 and 11-3. The size of the "between" symbol is shown in Figure 11-4.

When a profile tolerance applies all around the profile of a part, a circle is drawn at the bend of the leader attached to the feature control frame. The diameter of the circle is equal to the letter height. See Fig. 11-4.

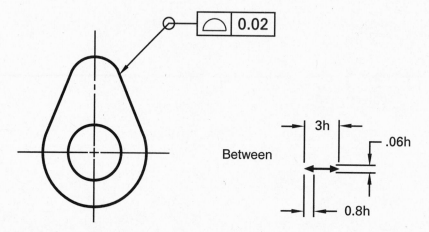

Fig. 11-4   An all around profile tolerance.

11-4   APPLICATION OF DATUMS

Profile controls can be applied with or without datums. When used without datums, the profile control can control form and size. Figures 11–4 and 11–5 show examples of profile as a form control. When applied with datums, the profile control can control size, form, orientation, and position. In these cases the limits of size rule does not apply.

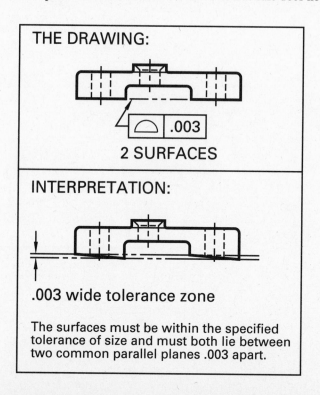

Fig. 11-5   Coplanarity controlled by a profile of a surface tolerance.

11-5   PROFILE TOLERANCE FOR COPLANAR SURFACES

Flatness is a control for a single plane surface. If a control is required for the flatness of two or more surfaces (coplanarity), profile of a surface is used. The profile control in Fig. 11-5 specifies a three-dimensional tolerance zone encompassing both surfaces. The two surfaces are considered one interrupted plane and are not related to any other feature of the part. Datums can be used in this type of application if interrelationship with other features is necessary.

11-6   PROFILE TOLERANCE FOR TAPERED SHAFTS

A tapered shaft can have its circular elements controlled with circularity and its straight elements controlled with straightness. The only control available to control both circularity and straightness of a tapered shaft is profile of a surface. This control can be a form control if it is applied without a datum or a related control if it is applied with a datum. The example in Fig. 11-6 illustrates a three-dimensional tolerance zone bilaterally applied around a basically defined tapered surface relative to a different diameter.

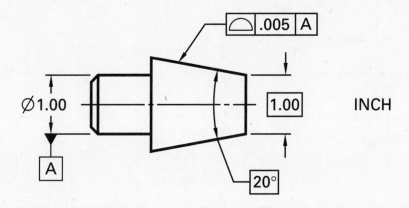

Fig. 11-6   Profile of a tapered shaft.

11-7   COMPOSITE PROFILE

In some instances the form and orientation requirements for the surface profile of an irregular feature of an object are more restrictive than the position requirement. In this case composite profile may be used.

Figure 11-7 illustrates an example where the position of a basically defined surface is controlled within a 1.0-mm tolerance zone relative to datums A, B, and C. A smaller tolerance zone of 0.3 mm is required for the form and orientation relative only to the A datum. The smaller 0.3-mm tolerance zone must be perpendicular to the A datum, but it is allowed to "float" relative to the B and C datums within the larger 1.0-mm tolerance zone. The actual surface must fit within both tolerance zones.

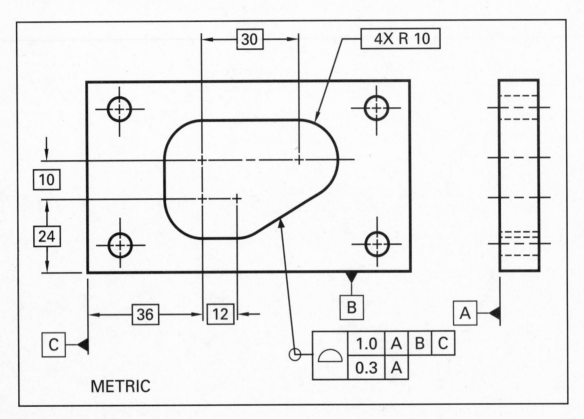

Fig. 11-7   Composite profile.

CHAPTER 12

Parallelism

12-1 DEFINITION

Two surfaces are *parallel* if every point on each surface is the same distance from the other surface. *Parallelism error* is the measure of the amount that two surfaces deviate from this condition. Parallelism can also apply to circular features, such as cylinders and cones, in which case the *axes* are parallel. For example, a cylinder may be parallel to another cylinder or to a flat surface.

12-2 PARALLELISM TOLERANCE ZONE

The tolerance zone for parallelism can take one of three forms, depending on the form of the controlled feature and the datum.

1. Whether the datum is a flat surface or a circular feature, if the controlled feature is a flat surface, the tolerance zone will be the space between two imaginary planes, perfectly parallel to the datum, within which the flat surface must lie (see Fig. 12-1).

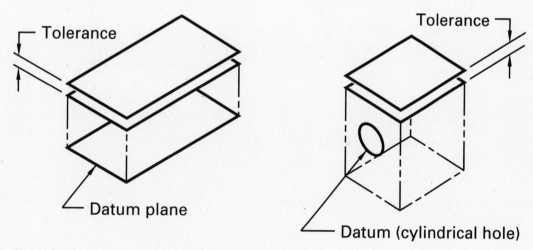

Fig. 12-1 Shape of tolerance zone when controlled feature is a flat surface.

2. If both the datum and the feature being controlled are circular features, such as cylinders and cones, the tolerance zone will be an imaginary cylinder whose axis is perfectly parallel with the axis of the datum. The axis of the controlled feature must lie within that cylinder. The tolerance zone is a cylinder because the axes of the two features must be parallel in all directions (see Fig. 12-2). In this case the tolerance is specified in the feature control frame as a diameter. (See, for example, Figs. 12-5 and 12-6).

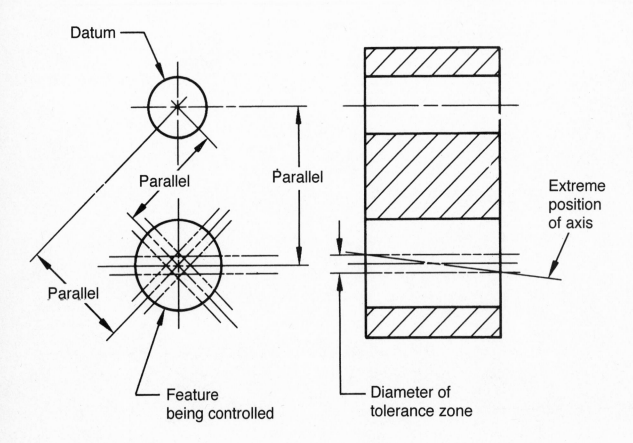

Fig. 12-2   Shape of tolerance zone when both controlled feature and datum are circular.

3. A less common form of the tolerance zone is the space between two imaginary parallel lines perfectly parallel to the datum. Usually the whole surface must be parallel to a datum, but in this case each element must be parallel to the datum. Below the feature control frame the note, "EACH ELEMENT" is written. The number in the feature control frame specifies the distance between two lines.

12-3   SPECIFYING PARALLELISM TOLERANCE

Examples of how to specify parallelism tolerance are given in Figs. 12-3 through 12-6. Because parallelism is a relationship characteristic, a datum feature symbol is always required. When features of size are involved, such as holes, the parallelism tolerance is understood to apply regardless of feature size (RFS). The limits of size rule applies to all size features that have their parrellelism controlled.

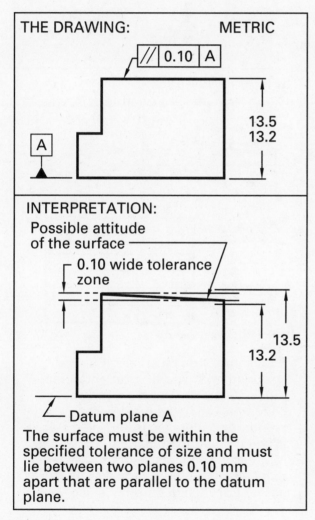

Fig. 12-3   Specifying parallelism of two flat surfaces.

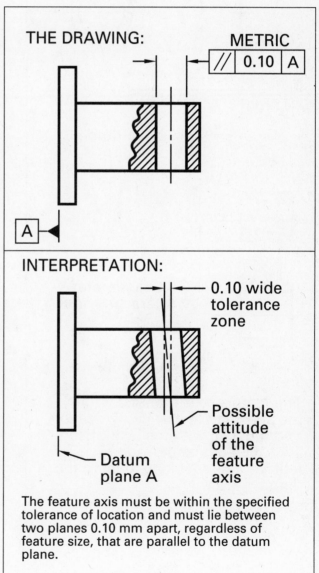

Fig. 12-4   Specifying parallelism of a circular feature and a flat surface.

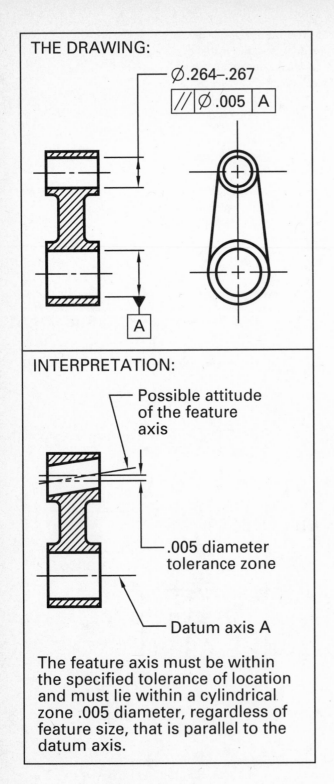

Fig. 12-5 Specifying parallelism of two circular features—both features RFS.

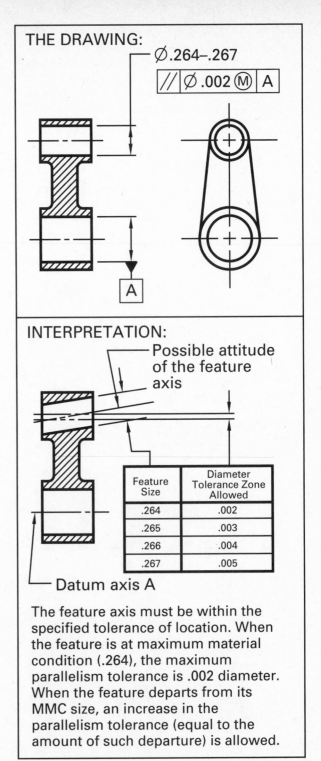

Fig. 12-6 Specifying parallelism of two circular features—controlled feature MMC.

Figure 12-3 shows an example of parallelism applied to a plane surface. The top surface of the object is controlled to the bottom surface within 0.1 millimeter. The top surface must meet the requirements of size and parallelism.

The axis of the hole in Figure 12-4 must be parallel to the left side of the object. Usually, when a hole is controlled with parallelism, a cylinder tolerance zone is used. In this case the tolerance zone is the space between two parallel planes.

The desire to have two holes parallel as shown in Figure 12-5 is a very common situation. The axis of the controlled hole (top) must be parallel to the axis of the datum hole (bottom). Measurements are obtained when the tolerance and datum feature are regardless of feature size. The distance between the holes may be a basic dimension or a toleranced dimension. The basic dimension application specifies an exact location for the tolerance zone. If a toleranced dimension is used, the axis of the hole may "float" within the size tolerance and meet the parallelism requirement.

Figure 12-6 illustrates what happens to the tolerance size when the tolerance is applied at maximum material condition and the feature is not at MMC. The parallelism tolerance is allowed to increase as the hole increases in size.

CHAPTER 13

Perpendicularity

13-1 DEFINITION

Two plane surfaces or straight lines are *perpendicular* when they are at a right angle (90°) to each other. They do not necessarily have to intersect to be perpendicular. For example, the wall in one building can be perpendicular to the floor of the next building. *Perpendicularity error* is any deviation in an actual part from a perfect right angle.

13-2 PERPENDICULARITY TOLERANCE ZONE

The tolerance zone can take any one of three forms, depending on the form of the controlled feature and the datum.

1. The most common form of the tolerance zone is the space between two imaginary parallel planes perfectly perpendicular to the datum. This occurs under the following conditions:

   a. The controlled feature and the datum are both flat surfaces. The controlled surface must be within the parallel planes (see Fig. 13-1).

   b. The datum is a flat surface and the controlled feature is a slot or tab. In this case the median plane (an imaginary plane in the center) of the slot or tab must be between the parallel planes (see Fig. 13-2).

   c. The controlled feature and the datum are both circular features, in which case the axis of the controlled feature must be between the parallel planes (see Fig. 13-3).

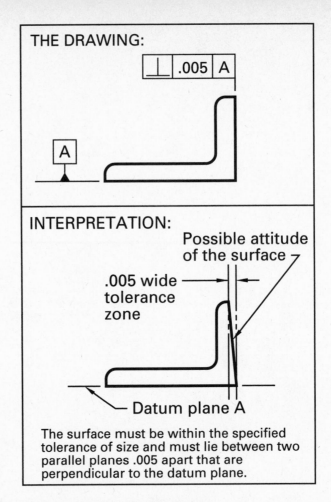

THE DRAWING:

⊥ | .005 | A

A

INTERPRETATION:

Possible attitude of the surface

.005 wide tolerance zone

Datum plane A

The surface must be within the specified tolerance of size and must lie between two parallel planes .005 apart that are perpendicular to the datum plane.

Fig. 13-1   Specifying perpendicularity of two flat surfaces.

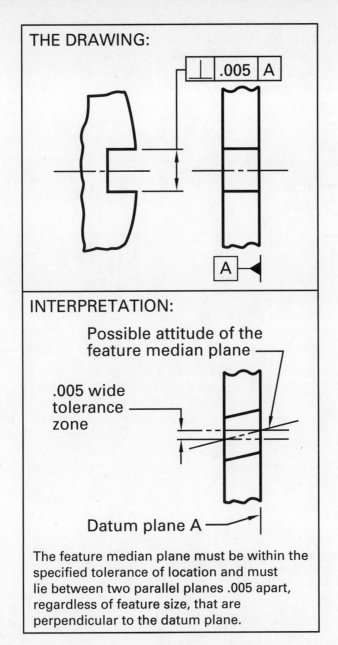

THE DRAWING:

⊥ | .005 | A

A

INTERPRETATION:

Possible attitude of the feature median plane

.005 wide tolerance zone

Datum plane A

The feature median plane must be within the specified tolerance of location and must lie between two parallel planes .005 apart, regardless of feature size, that are perpendicular to the datum plane.

Fig. 13-2 Specifying perpendicularity for a median plane—the center of a slot.

2. The tolerance zone can be in the form of an imaginary cylinder perfectly perpendicular to the datum, which occurs when the datum is a flat surface and the controlled feature is a circular feature (see Figs. 13-4 and 13-5). The axis of the circular feature must be within the imaginary cylinder. The tolerance zone is a diameter rather than a width because perpendicularity applies in all directions. An infinite number of parallel planes in all directions centered on an axis forms a cylinder. Note that the tolerance is specified as a diameter in the feature control frame.

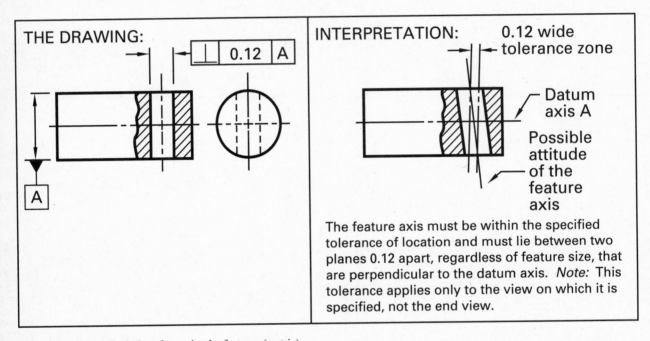

THE DRAWING:

INTERPRETATION: 0.12 wide tolerance zone

Datum axis A

Possible attitude of the feature axis

The feature axis must be within the specified tolerance of location and must lie between two planes 0.12 apart, regardless of feature size, that are perpendicular to the datum axis. *Note:* This tolerance applies only to the view on which it is specified, not the end view.

Fig. 13-3   Perpendicularity of two circular features (metric).

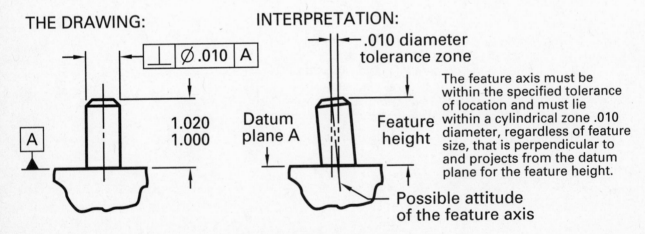

THE DRAWING:

INTERPRETATION: .010 diameter tolerance zone

1.020
1.000

Datum plane A   Feature height

The feature axis must be within the specified tolerance of location and must lie within a cylindrical zone .010 diameter, regardless of feature size, that is perpendicular to and projects from the datum plane for the feature height.

Possible attitude of the feature axis

Fig. 13-4   Specifying perpendicularity of a circular feature relative to a flat surface.

THE DRAWING:

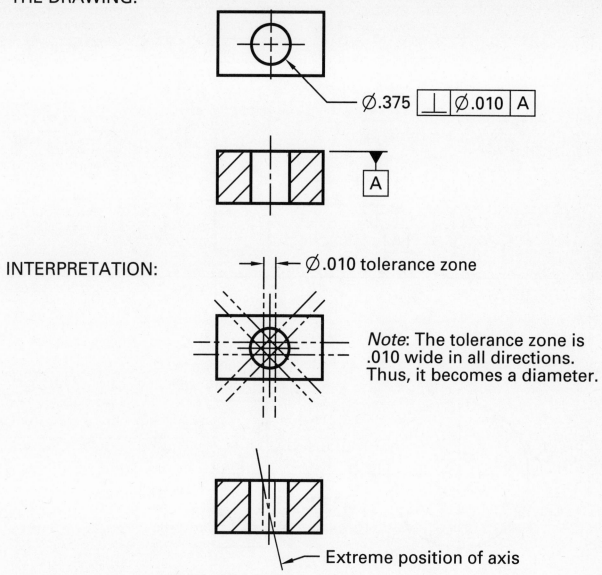

INTERPRETATION:

Ø.010 tolerance zone

Note: The tolerance zone is .010 wide in all directions. Thus, it becomes a diameter.

Extreme position of axis

Fig. 13-5 For a cylinder perpendicular to a flat surface, the tolerance zone is a diameter because the perpendicularity applies in all directions from the cylinder axis.

3. The least common form of the tolerance zone is an area between two imaginary parallel lines perfectly perpendicular to the datum. Figure 13-6 is an example where the individual lines of a flat surface are perpendicular to a cylindrical size feature. The whole surface may not be perpendicular, but each radial element must be perpendicular to the datum. Below the feature control frame, the note "EACH RADIAL ELEMENT" is added. Each radial element of the surface must be between the two parallel lines that make up the tolerance zone.

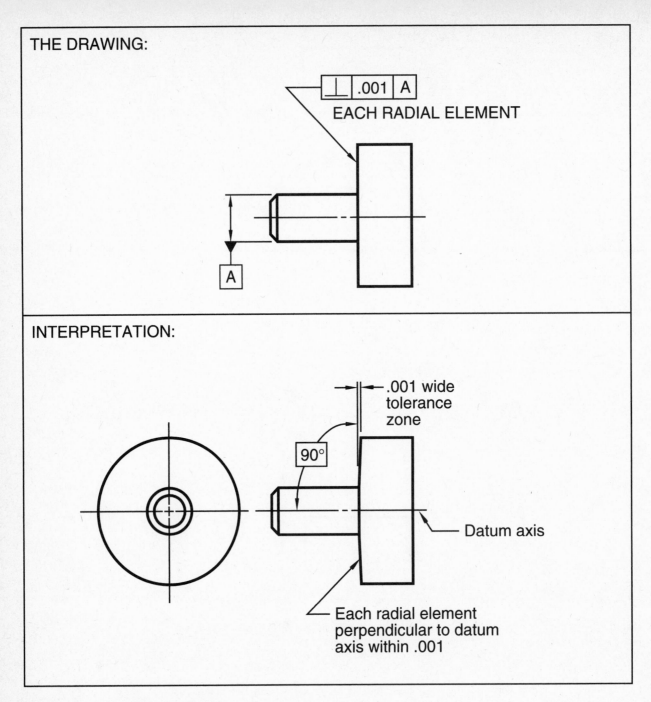

THE DRAWING:

⊥ .001 A

EACH RADIAL ELEMENT

A

INTERPRETATION:

.001 wide
tolerance
zone

90°

Datum axis

Each radial element
perpendicular to datum
axis within .001

Fig. 13-6   Specifying radial perpendicularity.

13-3   SPECIFYING PERPENDICULARITY TOLERANCE

Figures 13-1 through 13-6 provide examples of how to specify perpendicularity on drawings. Perpendicularity is a relationship characteristic, so a datum is always required. Note that when the datum is a symmetrical size feature, the datum feature symbol is never placed on the center line; it is always given with the size dimension (see Fig. 13-3).

13-4   USE OF MODIFIERS

Unless specified otherwise in the feature control frame, perpendicularity applies regardless of feature size. When necessary, the tolerance applies at the maximum material condition—Fig. 13-7 is an example. Two mating parts are shown, one having a slot and the other a tab. In the worst tolerance condition, the tab must still fit into the slot.

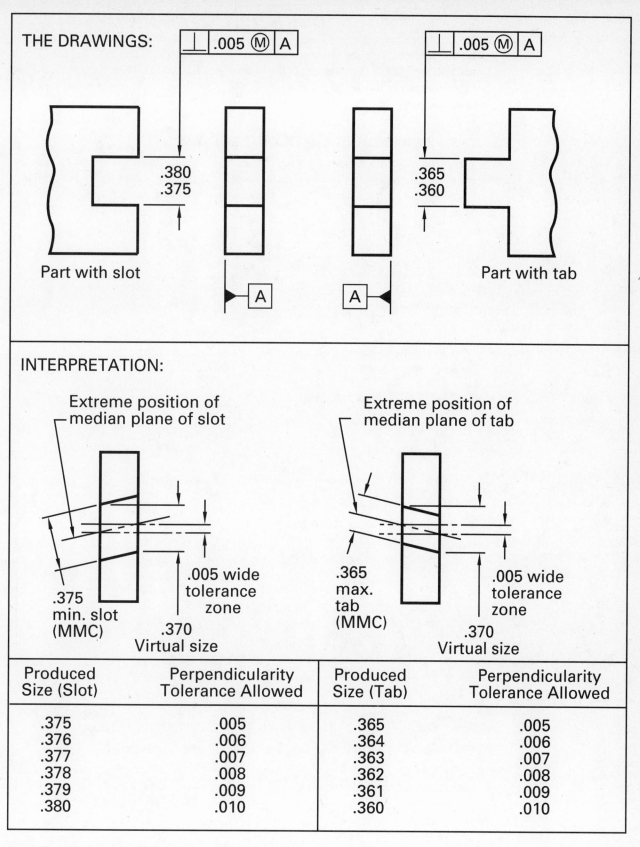

Fig. 13-7 Specifying perpendicularity at MMC—mating parts.

In Fig. 13-7, p. 67, the combination of the maximum material condition and the geometric tolerance create virtual conditions for the slot and the tab. The virtual conditions are used to determine the clearance between the mating parts. When the slot is at .375, it can be tilted by .005. The virtual condition for the slot is .370 (.375 − .005 = .370). The virtual condition for the tab is also .370 (.365 + .005 = .370). In the worst case, the clearance would be .000.

13-5   PERPENDICULARITY AND FLATNESS

A perpendicularity tolerance for a flat surface also controls flatness when no flatness tolerance is specified. Every point on the actual surface must be within the perpendicularity tolerance zone. Therefore, the out-of-flatness cannot exceed the perpendicularity tolerance.

CHAPTER 14

Angularity

14-1 INTRODUCTION

Angularity tolerance specifies the permissible error in an angle in a way that is different from the ordinary size tolerance attached to an angular dimension. A tolerance expressed in degrees results in a fan-shaped tolerance zone, as shown in Fig. 14-1.

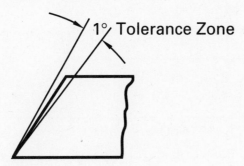

Fig. 14-1 Angular size tolerance.

The tolerance is actually zero at the vertex and increases along the length of the angular surface. The increase for a 1° tolerance zone is about .017 inch per inch of length. For a 10-inch length (254 mm), it will be more than .170 inch (4.32 mm). This may not be acceptable in the design. Instead, a tolerance zone of uniform width for the entire surface may be required. This requirement can be met by specifying the tolerance, not in degrees, but in units of an inch or in parts of a millimeter, and this is exactly the function of a geometric angularity tolerance. An angularity tolerance specifies the uniform width of a tolerance zone along the entire surface.

14-2 ANGULARITY TOLERANCE ZONE

The angularity tolerance zone can take one of three different forms:

1. Two parallel planes at a basic angle to a datum plane or axis. The controlled surface, center plane, or axis must be within the two planes.
2. A cylinder at a basic angle to one or more datum planes or axes. The controlled axis must be within the cylinder.
3. Two parallel lines at a basic angle to a datum plane or axis. The line elements of the controlled surface must be within the two lines.

Everything that has been said about angularity so far sounds much like perpendicularity, and it is. Perpendicularity is a special case of angularity for an angle of 90°. Angularity is the general case for all other angles.

14-3   SPECIFYING ANGULARITY TOLERANCE

The angular dimension giving the required angle must be *basic* (without tolerance) because the tolerance will be specified in the feature control frame. The feature control frame is directed by means of a leader to the line representing the controlled feature or to an extension from that line (see Fig. 14-2). Another method is to draw the feature control frame with a corner of it touching the extension line (see Fig. 14-3). When the feature being controlled is a size feature, the feature control frame can be drawn to the right of or below the size dimension, as shown in Fig. 14-4. Because angularity is a relationship characteristic, a datum is always required. Note that the feature control frame and datum feature symbol are never placed on the center line representing the axis or median plane of the feature.

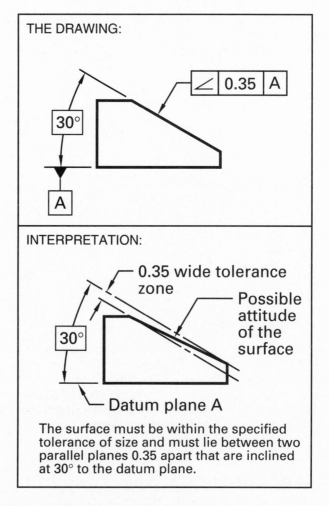

Fig. 14-2   Specifying angularity for a flat surface (metric).

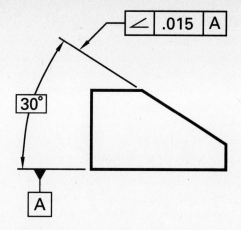

Fig. 14-3 Specifying angularity—alternate placement of the feature control frame.

THE DRAWING:

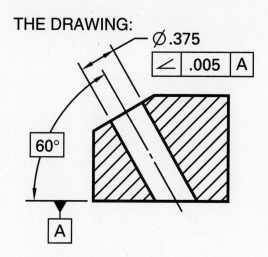

INTERPRETATION:

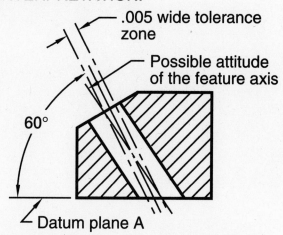

The feature axis must be within the specified tolerance of location and must lie between two parallel planes .005 apart that are inclined at 60° to the datum plane.

Fig. 14-4 Specifying angularity for a size feature.

14-4   USE OF MODIFIERS

Unless otherwise specified in the feature control frame, angularity applies regardless of feature size. When dealing with size features, the tolerance may apply at the maximum material condition. A tab at some angle, for example, may have to fit into a slot in another part. However, this is best specified using a *position* tolerance with a circle M modifier, which is explained in Chapter 18. (Figure 18-14 is an example of a position tolerance used to control angularity.)

14-5   ANGULARITY AND FLATNESS

As with perpendicularity and parallelism, angularity also controls flatness of plane surfaces when no flatness tolerance is specified. Every point on the actual surface must be within the angularity tolerance zone. Therefore, the out-of-flatness cannot exceed the angularity tolerance.

CHAPTER 15

Concentricity

15-1 COAXIALITY

Two cylinders are *coaxial* if they have the same axis or if their axes are on one line. See the examples on the next page in Fig. 15-1. The word *coaxiality* is used to describe the general case where the axes of different features are in line. There is no geometric characteristic symbol for coaxiality. Instead, any one of three symbols that define special cases is used: concentricity, runout, or position. This chapter will examine concentricity. Chapter 16 will deal with runout, and Chapters 17, 18, and 19 will deal with position.

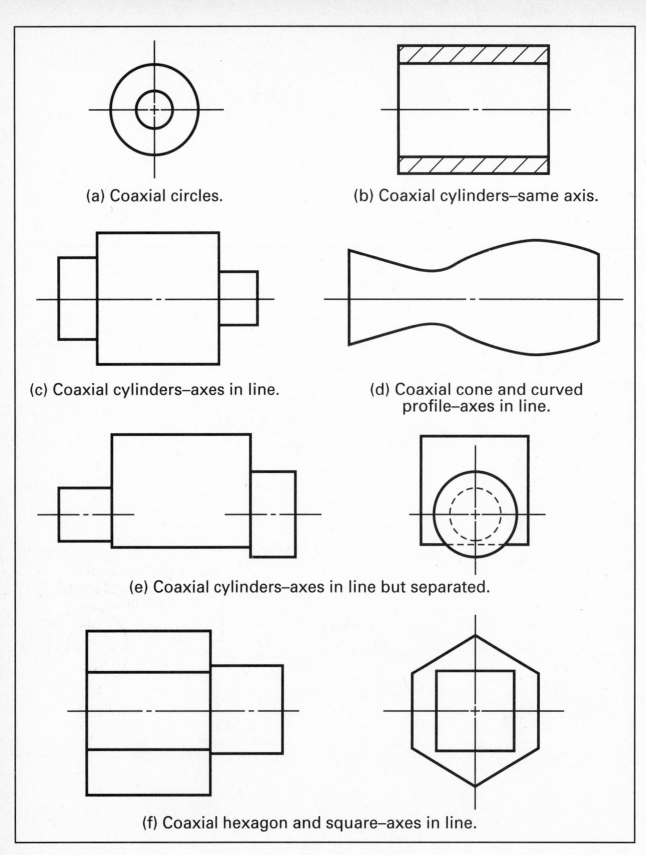

(a) Coaxial circles.

(b) Coaxial cylinders–same axis.

(c) Coaxial cylinders–axes in line.

(d) Coaxial cone and curved profile–axes in line.

(e) Coaxial cylinders–axes in line but separated.

(f) Coaxial hexagon and square–axes in line.

Fig. 15-1 Examples of coaxiality.

15-2   CONCENTRICITY TOLERANCE

Concentricity error is the amount by which the axes of two cylinders are out of line. However, it is not measured as the distance between the axes. Error in concentricity is sometimes referred to as *eccentricity* (literally "off-centeredness"), but this term is generally used in connection with circular shapes that are off-center by design. Concentricity is more restricted in its use than the other coaxiality characteristics and therefore it is specified least often.

15-3   CONCENTRICITY TOLERANCE ZONE

The form of the tolerance zone is an imaginary cylinder about the exact axis of the datum. The diameter of this cylinder is equal to the specified tolerance. The axis of the controlled feature must lie within the imaginary cylinder—see Figs. 15-2 and 15-3. Notice that the axis of the controlled feature can be anywhere within the tolerance zone. Therefore, it can be slanted relative to the datum axis. This is one reason why concentricity error is not measured as the distance between the datum axis and the feature axis. If the feature axis were slanted, it would be parallelism error, not concentricity error. However, the slant is treated as an error in concentricity.

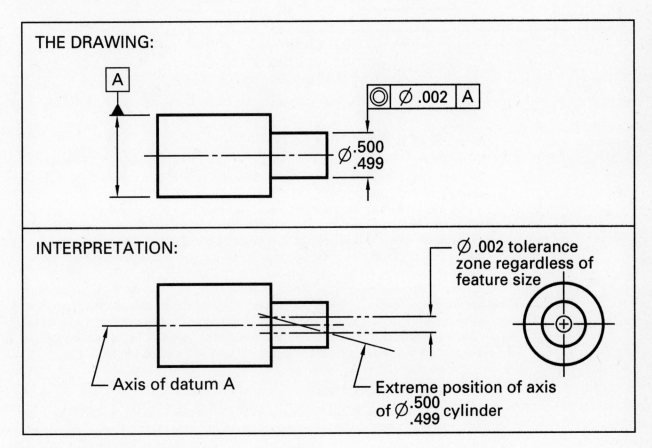

Fig. 15-2   Specifying concentricity—datum is a single cylinder.

THE DRAWING:

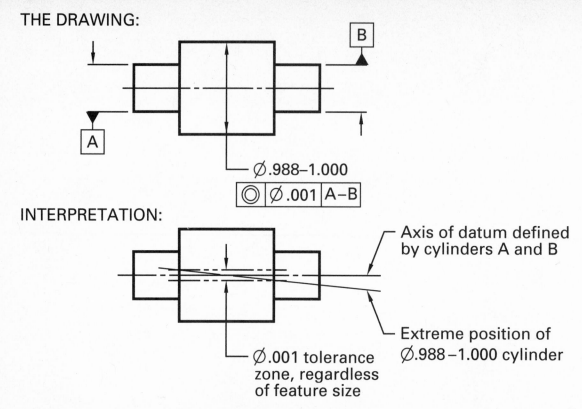

Fig. 15-3   Specifying concentricity—datum defined by two separated cylinders.

15-4   SPECIFYING CONCENTRICITY

Concentricity is specified as shown in Figs. 15-2 and 15-3. Since the form of the concentricity tolerance zone is a cylinder, the tolerance is specified as a diameter. Place the diameter symbol ($\varnothing$) *before* the tolerance.

A datum is always necessary because concentricity is a relationship characteristic, like parallelism and perpendicularity. Select as datums surfaces that are functional in the use of the part. In a cylindrical part, this might be a diameter that fits into a bearing or mating part, or it might be a flat face that locates the part in an assembly. Lathe centers are never used because they have no function in the use of the part.

The datum may be a single cylinder, as in Fig. 15-2, or it may be two separated cylinders used as one datum, as in Fig. 15-3. In this situation the object is rested on both cylinders for concentricity testing of the controlled diameter. The two cylinders comprise one datum. This situation is indicated in the feature control frame by connecting the two datum letters with a dash rather than separating them with a vertical line.

15-5   USE OF MODIFIERS

Concentricity tolerance always applies regardless of the size of the controlled feature and the datum (RFS). If a fit at the maximum material condition (MMC) is required, position tolerance is used.

15-6   TESTING FOR CONCENTRICITY ERROR

Because it is impossible to be inside an object and measure the distance between the axes of cylinders, concentricity must be measured from the outside. This detailed and time-consuming process involves taking differential measurements around the shaft at any cross-section and comparing each measurement to make sure it fits within the acceptable cylinder (see Fig. 15-4). All differential measurements from any cross-section must fit in the cylinder.

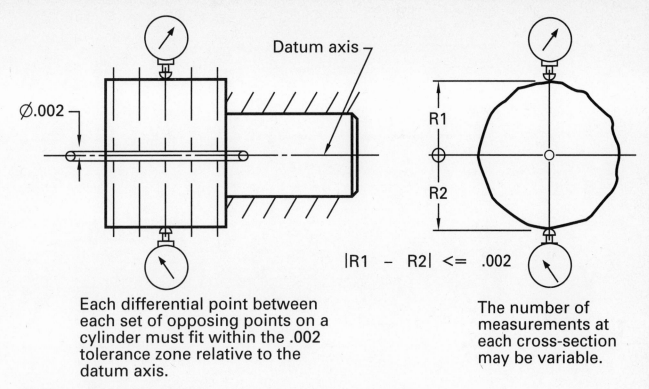

Fig. 15-4   Concentricity measurement.

15-7   SELECTION OF PROPER CONTROL FOR COAXIALITY

Use concentricity when the following conditions apply:

1. Coaxiality must be controlled independently of surface errors.
2. The desired coaxiality tolerance must be held regardless of feature and datum size (RFS).

Use runout when the following conditions apply:

1. Surface errors can be included with coaxiality error.
2. The desired coaxiality tolerance must be held regardless of feature and datum size (RFS).

Use position tolerance when the following conditions apply:

1. Surface errors can be included with coaxiality error.
2. The coaxial features must assemble with another part having corresponding coaxial features, and additional coaxial tolerance may be allowed when the feature and/or the datum are not at the maximum material condition (see Chapter 19).

CHAPTER 16

Runout

16-1 INTRODUCTION

Runout is any deviation of a surface from perfect form that can be detected by rotating the part about an axis. Runout tolerance is the maximum deviation allowed. It is a composite tolerance including errors in circularity, straightness, perpendicularity, and coaxiality. All these geometric errors can be read as runout if they are detected by rotating the part about an axis. Runout can be applied to any surface generated around an axis, such as cylinders, cones, and curved profiles. It is also applied to flat faces perpendicular to an axis.

16-2 CIRCULAR AND TOTAL RUNOUT

There are two types of runout control: circular runout and total runout. In circular runout, the deviation of each circular element is controlled; there is no control over elements in any other direction. Total runout controls the deviation of all elements of a surface—circular or straight. The entire surface is controlled. Circular runout is less expensive to measure and is adequate for most design functions. Therefore, it is more commonly specified than total runout.

16-3 RUNOUT TOLERANCE ZONE

The shape of the tolerance zone is different for circular and total runout, and for each type of control it can take any of four forms, depending on the shape of the feature being controlled. All these forms are tabulated and illustrated in Figs. 16-1 and 16-2. Notice that for circular runout the tolerance is the distance between theoretical *elements,* while for total runout it is the distance between theoretical *surfaces.*

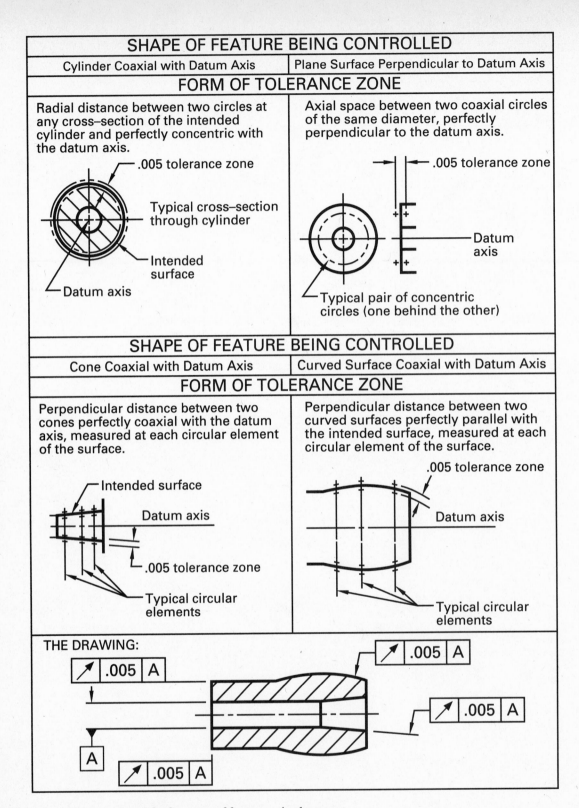

Fig. 16-1 Form of tolerance zone for four types of features—circular runout.

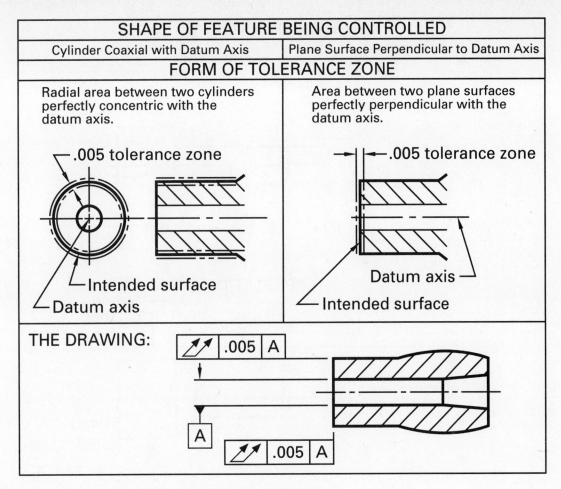

Fig. 16-2   Form of tolerance zone for two types of features—total runout.

16-4   SPECIFYING RUNOUT

Runout tolerance is specified in the same way as other geometric tolerances that express a relationship of one feature to another. One or more datums are identified on the drawing and specified in the feature control frame. Examples of both types of runout control are shown in Figs. 16-3 through 16-8. The double arrow symbol for total runout illustrates the fact that the surface must be tested in two directions (circular elements and straight elements).

THE DRAWING:

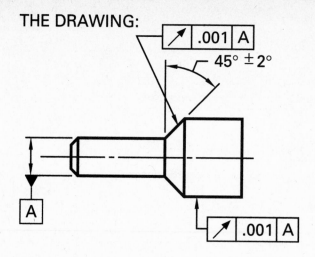

INTERPRETATION:

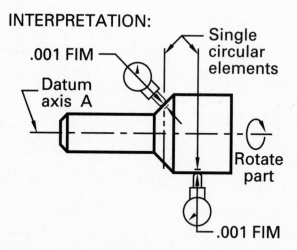

Fig. 16-3 Circular runout—datum is a single cylinder.

THE DRAWING:

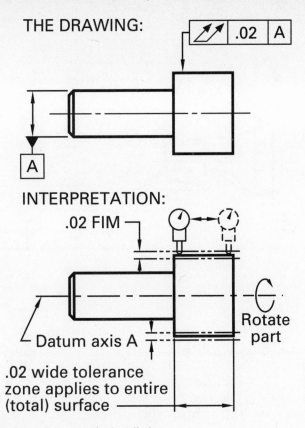

INTERPRETATION:

Fig. 16-4   Total runout—datum is a single cylinder.

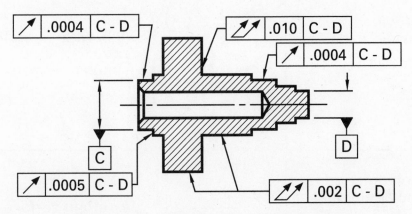

Fig. 16-5   Circular and total runout—datum defined by two separated cylinders.

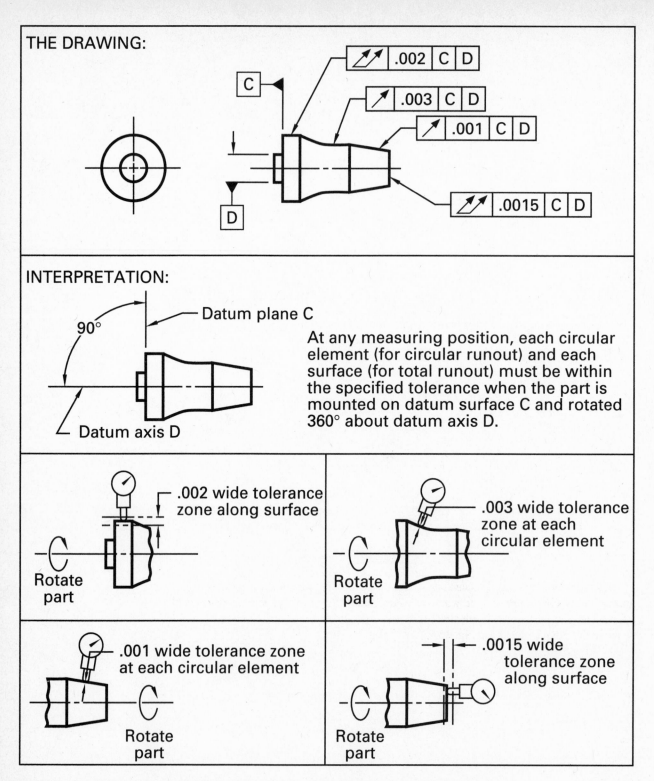

Fig. 16-6 Specifying and testing circular and total runout—a flat surface and a cylinder used as datums.

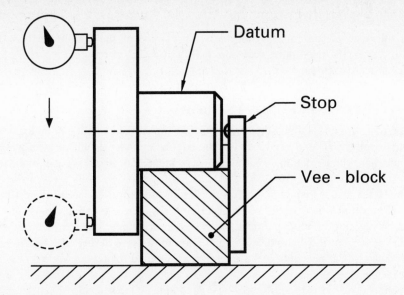

Fig. 16-7   A flat-face datum held against tooling for testing runout.

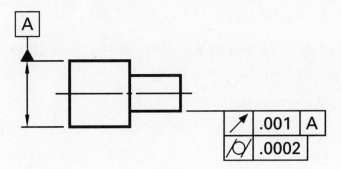

Fig. 16-8   Combined runout and cylindricity tolerance.

As with most geometric characteristics, runout can be combined with other tolerances. For example, a diameter with a .05-mm runout tolerance may have to be held cylindrical within .02 mm. To illustrate this situation, attach a cylindricity feature control frame to the runout frame (see Fig. 16-8).

16-5   APPLICATION OF DATUMS

As in concentricity tolerance, select as datums surfaces that are important in the function of the part. In a cylindrical part, this might be a diameter that fits into a bearing or other mating part, or it might be a flat face that locates the part in an assembly. Lathe centers should not be used because they have no function in the use of the part.

The datum can be a single cylinder, as in Figs. 16-3 and 16-4, or it may be two separated cylinders used as one datum, as in Fig. 16-5. In this situation the object is rested on both cylinders for testing runout of the controlled feature. The two cylinders comprise one datum. This situation is indicated in the feature control frame by connecting the two datum letters with a dash rather than separating them with a vertical line.

When cylindrical parts are relatively large in diameter and short in length, specify a flat face as a datum in addition to a cylindrical surface (see the parts in Figs. 16-6 and 16-7). In this case the part would tilt if laid in a vee-block for testing, so it is steadied by holding the flat face against the appropriate tooling perpendicular to the vee.

16-6   USE OF MODIFIERS

Runout tolerance always applies regardless of the size of the feature and the datum (RFS). If a fit at the maximum material condition (MMC) is required, specify position tolerance instead of runout.

16-7   TESTING FOR RUNOUT ERROR

Runout error is commonly measured by contacting the surface with a dial indicator while the part is rotated 360° in a vee-block (for an external cylinder datum) or over a mandrel (for an internal cylinder datum). The full indicator movement (FIM) is equal to the runout error.

When a flat face is inspected, place a stop against the opposite face of the part to prevent axial movement that would also register on the indicator (see Fig. 16-7). In testing for circular runout on both cylindrical and plane surfaces, only circular elements are inspected. The body of the dial indicator is not allowed to move during each rotation of the part. Enough circular elements are inspected to ensure that the readings are typical of all circular elements (see Fig. 16-3).

When total runout is tested, several circular elements are inspected; then several longitudinal elements (on cylinders) or radial elements (on flat faces) are inspected. The latter inspections are done by moving the dial indicator longitudinally or radially as required while the part is held still. After each inspection the part is rotated to a new position and another element is tested (see Fig. 16-4).

16-8   SELECTION OF PROPER CONTROL FOR COAXIALITY

Use runout when the following conditions apply:

1. Surface errors can be included with coaxiality error.
2. The desired coaxiality tolerance must be held regardless of feature and datum size (RFS).

Use concentricity when the following conditions apply:

1. Coaxiality must be controlled independently of surface errors.
2. The desired coaxiality tolerance must be held regardless of feature and datum size (RFS).

Use position tolerance when the following conditions apply:

1. Surface errors can be included with coaxiality error.
2. The coaxial features must assemble with another part having corresponding coaxial features, and additional coaxiality tolerance may be allowed when the feature and/or the datum are not at the maximum material condition (see Chapter 19).

CHAPTER 17

Position Tolerance—General

17-1 DEFINITION

The position of a feature is the theoretically exact location of its axis or center plane from the feature or features from which it is dimensioned. Any deviation from the theoretically exact position is known as *position error. Position* is the total permissible error in the location of a feature relative to one or more features of a part.

17-2 POSITION TOLERANCE

Position is a composite tolerance. It may be composed of errors in parallelism and/or perpendicularity, in addition to the mislocation of an axis or center plane from an exact location. Position can be used to control the location and orientation of internal or external features of size. Because of this flexibility, it is the most used of all the geometric controls.

17-3 POSITION TOLERANCE ZONE

The form of the tolerance zone for position tolerance depends on whether the feature is located on an axis or a center plane. For a feature located by its axis, such as a round hole or pin, the tolerance zone is the space within an imaginary cylinder perfectly parallel, perpendicular, or concentric to one or more datums and centered at the theoretically exact position of the axis (see Fig. 17-1). The diameter of the cylinder is equal to the specified tolerance; its length is equal to the length of the feature. The axis of the feature as produced must lie entirely within the cylinder. The tolerance value given in the feature control frame is always specified as a diameter. (For example, in Fig. 17-6, it is ⌀.007.)

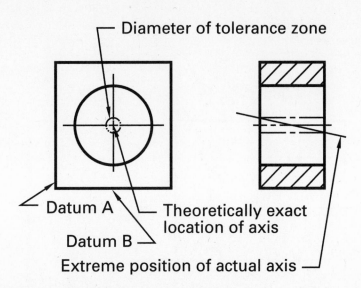

Fig. 17-1 Position tolerance zone for a feature located by its axis.

For a feature located by its center plane, such as a slot or tab, the tolerance zone is the space between two imaginary planes perfectly parallel or perpendicular to and equidistant from the true position of the center plane (see Fig. 17-2). The distance between the imaginary planes is equal to the specified tolerance. The area of the tolerance zone is equal to the area of the feature. The center plane of the feature as produced must lie between the imaginary planes. The tolerance specified in the feature control frame is the *total* width of the tolerance and not the distance from the center plane of the datum to either side of the tolerance zone.

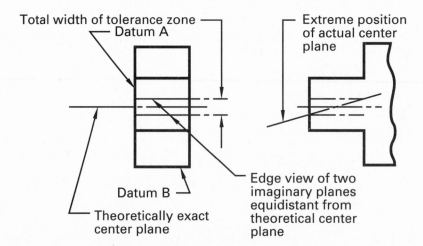

Fig. 17-2   Position tolerance zone for a feature located by its center plane.

17-4   COMPARISON OF COORDINATE TOLERANCING WITH POSITION TOLERANCING

In the past, patterns of features were dimensioned using coordinate tolerancing. The remainder of this chapter will be devoted to illustrating some of the limitations of this system and to comparing it with position tolerancing.

It is logically incorrect to explain the differences in the dimensioning systems using only one part. Understanding the consequences requires two mating parts. Part (a) in Fig. 17-3 has four holes with diameters of .375 and tolerances of +.005/−.000. Part (b) has four pins (may represent threaded fasteners) with diameters of .361 and a tolerance of +.000/−.005. The distance between the centers of the pins or holes is 2.000, with tolerances of +.007/−.007. For each part, this type of dimensioning results in a square tolerance zone of .007 per side (see Fig. 17-4). The centers of the features (holes and pins) must fall within this tolerance zone. In their assembly, each set of mating features will use one .007 square zone.

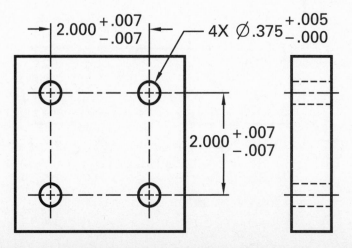

Fig. 17-3   Mating parts dimensioned by coordinates, Part A. *(continues)*

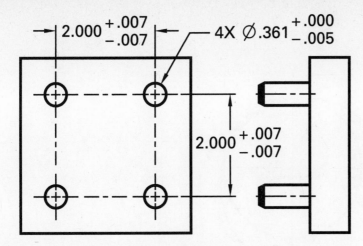

Fig. 17-3 Mating parts dimensioned by coordinates, Part B. *(continued)*

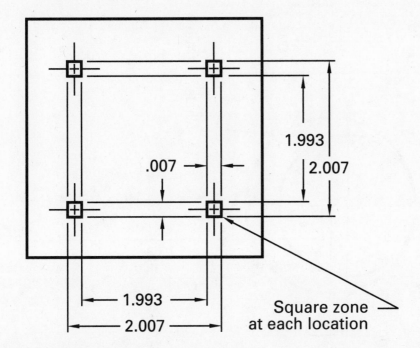

Fig. 17-4 Square tolerance zones.

Different center locations of two mating features are shown in Fig. 17-5. The farthest distance apart of the actual centers of the features on the horizontal axis is shown in part (a). The farthest distance apart on the vertical axis is shown in part (b). In each case the worst that can happen is a line-to-line fit (all of the clearance between the hole and pins is on one side). However, when the features are located in the opposite corners of the tolerance zone, as in part (c), there could be an interference (indicated by the cross-hatched area). This interference could mean that the parts will not assemble. This situation causes problems for the people who make and inspect parts. To eliminate any interference using coordinate tolerancing, the .007 tolerance must be reduced to .005 per side square zone (see Fig. 17-6). This reduction is accomplished by inscribing a square within the area of the .007 zone that won't allow interference to occur. The smaller .005 square zone is 36 percent smaller than the .007 diameter zone.

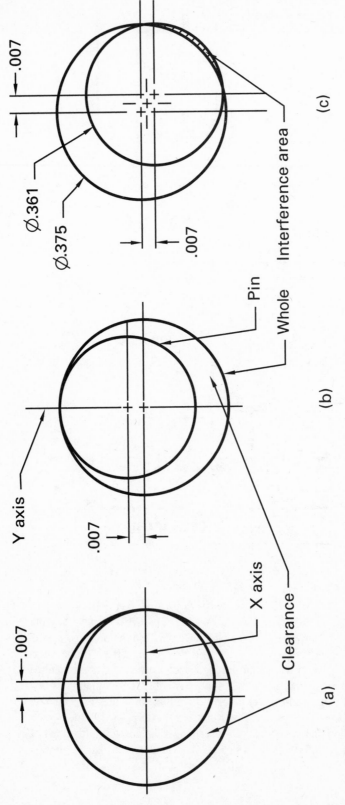

Fig. 17-5 Possible feature locations.

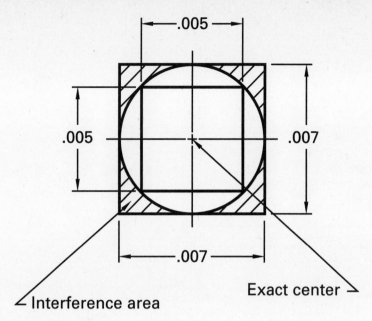

Fig. 17-6 Interference area.

17-5 SPECIFYING POSITION

Figure 17-7 shows part (b) from Fig. 17-3 with the holes controlled by a position tolerance. All the location dimensions are basic dimensions (without tolerance) because they
now give only the theoretically exact location of the holes. The tolerance is given in the
feature control frame. In this case, the largest functional tolerance zone is a .007 cylinder.

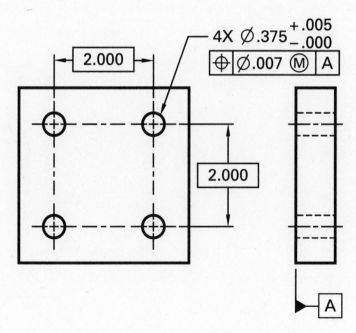

Fig. 17-7 Part (b) from Fig. 17-3 with the holes controlled by a position tolerance.

Three possible effects of different positions on the actual axis of a feature are shown in Fig. 17-8. Notice that the position tolerance also controls the perpendicularity of the holes relative to the surface of the part. The axis may be tilted by .007 within the thickness of the part.

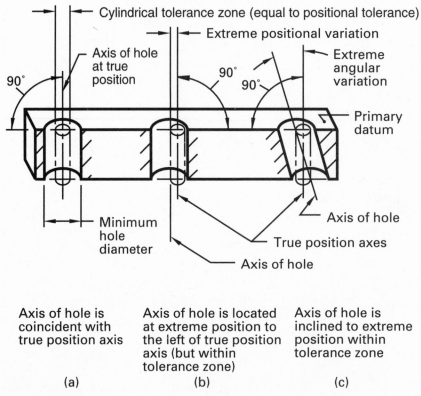

Axis of hole is coincident with true position axis

(a)

Axis of hole is located at extreme position to the left of true position axis (but within tolerance zone)

(b)

Axis of hole is inclined to extreme position within tolerance zone

(c)

Note that the length of the tolerance zone is equal to the length of the feature, unless otherwise specified on the drawing.

Fig. 17-8 Datum B has its own geometric form control.

17-6 USE OF MODIFIERS ON THE TOLERANCE

Position, like any other size feature control, is applied regardless of feature size. It can be applied at maximum material condition (MMC) or at least material condition (LMC).

Regardless of Feature Size

Regardless of feature size is the default condition for position. Regardless of the size of the controlled features, the stated tolerance will apply. This condition is most often applied when there is no clearance between the datum(s) and the mating part (see Fig. 17-9).

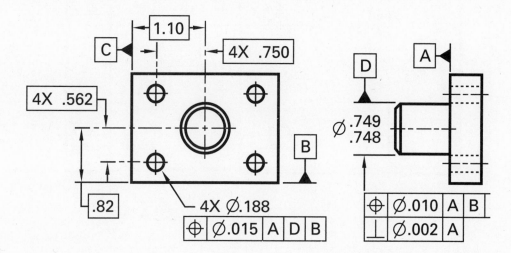

Fig. 17-9 Position regardless of feature size.

Maximum Material Condition

In Fig. 17-10, the tolerance has been modified with an MMC modifier, which means that the specified tolerance applies only when the feature is at MMC.

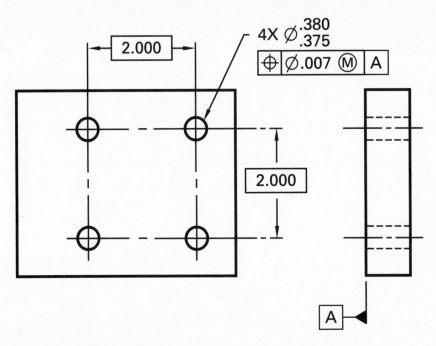

Fig. 17-10   Maximum material condition.

As the size departs from MMC toward LMC, the tolerance zone may increase in size. If one of the holes in Fig. 17-10 were produced at .378, the clearance between the hole and the pin would increase by .003 inches. This extra clearance, which is the difference between the actual size and the MMC size, is called a *bonus tolerance*. The following table could be made comparing various produced sizes with resulting position tolerances:

Actual Size	Position Tolerance
.375	.007
.376	.008
.377	.009
.378	.010
.379	.011
.380	.012

This increased tolerance zone size applies only to individual features independent of the other features in the pattern.

Least Material Condition

If the tolerance is applied to a hole at its largest size, or a pin is at its smallest size, LMC can be specified. Figure 17-11 illustrates an example where the surface of the hole is close to the edge of the part. If this distance must be controlled, the LMC modifier can be used. In the case of the part shown in Fig. 17-11, the edge of the hole will never be closer than 3.435 to the edge of the part because, as the hole gets smaller, the tolerance is allowed to increase, but the surface of the hole will not get any closer to the edge of the part.

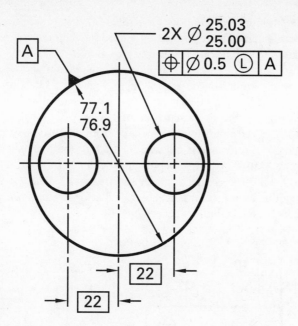

Fig. 17-11   An application of LMC to maintain a critical edge distance (metric).

17-7   DATUMS FOR POSITION

One or more datums should be specified with position. An example of two holes positioned to more than one datum is shown in Fig. 17-12. The datums in this case have geometric constraints on them. The B datum must be flat within 0.05 and perpendicular to datum A within 0.015. Datum C has similar requirements. Having such geometric constraints is called *qualifying the datums*. The datums must be acceptable before they are used to check the two holes. The holes are then checked against the three datums. The axes of the holes must be perpendicular to datum A while being located, and parallel, to datums B and C.

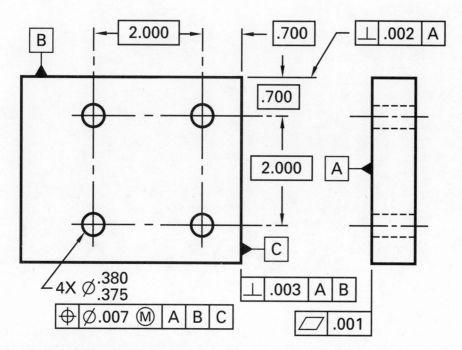

Fig. 17-12   Multiple datums and controls.

When a datum is a size feature, the condition of the datum and whether or not the datum has been controlled relative to another feature of the part must be determined. In

Fig. 17-9, the condition of the datum was regardless of feature size and it was not controlled to any other feature. The regardless of feature size application to a datum may be used for press-fit or threaded feature applications.

For most interchangeable assembly applications, the datum will be applied at MMC and it will be controlled relative to another feature. Figure 17-13 shows an example like the one in Fig. 17-9, but the four holes and the datum are controlled at MMC. In this case, the datum feature size applies at virtual condition, even though MMC is specified, because the datum is allowed to tilt by .002 inches before it is used as a datum. This does not mean that additional tolerance is allowed for each of the four holes. The four holes may shift as a group relative to the datum when the datum is not at MMC.

While controlled features are occasionally specified at LMC, datums are rarely specified at LMC. It is difficult to inspect parts controlled at LMC and usually does not represent an interchangeable fit condition.

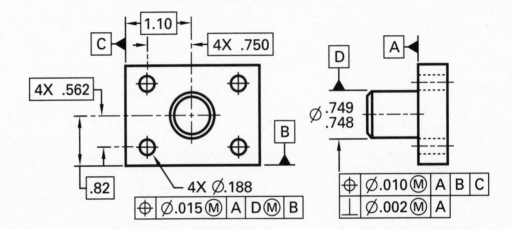

CHAPTER 18

Position Tolerance— Location Applications

18-1   MATING PARTS ASSEMBLED WITH FASTENERS

The advantage of position tolerance shows best when it is applied to mating parts fastened with bolts, screws, pins, and studs. Such parts may be held together in two different ways:

1. Floating fasteners (Fig. 18-1)—Both parts have plain cylindrical holes (clearance holes) that are larger than the fasteners. Since the fasteners are free to float within the clearance holes, assemblies of this kind are called *floating fasteners*.

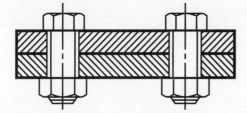

Fig. 18-1   Floating fastener assembly.

2. Fixed fasteners (Fig. 18-2)—One part has a clearance hole and the other has a tight-fitting hole that holds the fastener in a fixed position. The tight hole may be a press fit for a pin or it may be a tapped hole to accommodate a screw or stud. These types of assemblies are known as *fixed fasteners*.

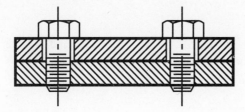

Fig. 18-2   Fixed fastener assembly.

In the design of mating parts assembled with fasteners, it is necessary to make decisions about the amount of clearance required in the holes and the position tolerance to be specified because the two are related. For floating fasteners, the tolerance is equal to the clearance; for fixed fasteners, the tolerance is equal to half the clearance.

The formulas given below will produce a metal-to-metal fit when the holes and fasteners are at MMC and the hole axes are at the extreme locations permitted by the position tolerance. When holes and fasteners as produced are not at their MMC, there will be more clearance—or additional position error can be allowed. The following symbols are used.

$$F = \text{Max Dia of } Fastener \text{ (MMC)}$$
$$H = \text{Min Dia of } Hole \text{ (MMC)}$$
$$T = \text{Dia of Position } Tolerance$$

18-2   ANALYSIS OF FLOATING FASTENERS

Once a decision has been made on the fastener size required, the hole size and position tolerance can be calculated by the following formulas.

$$H = F + T \qquad T = H - F$$

From the second formula, the tolerance (T) is equal to the clearance (H − F).

EXAMPLE:

Assume that the fastener in Fig. 18-1 is $\varnothing.250$ (MMC) and the clearance holes are $\varnothing.280$ minimum (MMC). What is the required position tolerance?

$$T = H - F$$
$$= .280 - .250$$
$$T = .030$$

In this example a position tolerance of $\varnothing.030$ is applied to *each* part. The total tolerance in both parts is .060. This tolerance can be divided unequally between the parts. The tolerance could be .040 on one part; in which case the tolerance on the other part would be .020.

18-3   ANALYSIS OF FIXED FASTENERS

The fixed fastener condition differs from floating fasteners because the fastener is fixed in one of the parts. It cannot change its location to accommodate position error. A fastener size is selected and then the hole size and position tolerance are calculated by the following formulas.

$$H = F + 2T \qquad T = \frac{H - F}{2}$$

In the second formula, H − F is equal to the clearance between the hole and the fastener. Thus, the position tolerance is equal to one half the clearance.

EXAMPLE:

Assume that the fastener in Fig. 18-2 is $\varnothing.250$ (MMC) and the clearance hole in the mating part is $\varnothing.280$ minimum (MMC). What is the required tolerance?

$$T = \frac{H - F}{2}$$
$$= \frac{.280 - .250}{2}$$
$$T = .015$$

In this example a position tolerance of $\varnothing.015$ is applied to each part. The total tolerance in both parts is .030. This tolerance can be divided unequally between the parts. The tolerance could be .020; in which case the tolerance on the other part would be .010.

18-4   PROJECTED TOLERANCE ZONE

Where the clearance hole is calculated by the fixed-fastener formula, the parts will sometimes not assemble because of position error that permits the axis of the fixed hole to be out of square with the surface. An example is shown in Fig. 18-3. If the upper part in Fig. 18-3 were thin, there may be no interference, and if the part were thicker, there would be even more interference. It is evident, therefore, that the height of the position tolerance

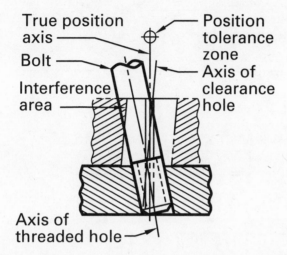

Fig. 18-3   Interference of mating part caused by out-of-squareness of tapped hole axis.

zone for the tapped hole should be based on the height of the mating part and not on the height of the tapped hole. The projected tolerance zone provides for this condition by transferring the tolerance zone from the part being controlled to the mating part (see Fig. 18-4).

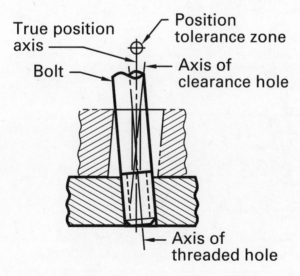

Fig. 18-4   Interference eliminated by making length of position tolerance equal to height of mating part.

Comparing this figure with Fig. 18-3, it is evident that the clearance hole size and the position tolerance have not been changed. The difference is that, in Fig. 18-4, the tolerance zone for the tapped hole is not within the tapped hole but is projected upward to a height equal to the height of the mating part. The mating part, of course, has its own position tolerance for the clearance hole. The two tolerance zones now coincide and this results in less perpendicularity error. The parts will now assemble properly when normal fixed-fastener clearance hole sizes are used.

An example of a projected tolerance zone specification is given in Fig. 18-5. The drawing of the mating part containing the clearance holes will specify the same position tolerance without a projected tolerance zone.

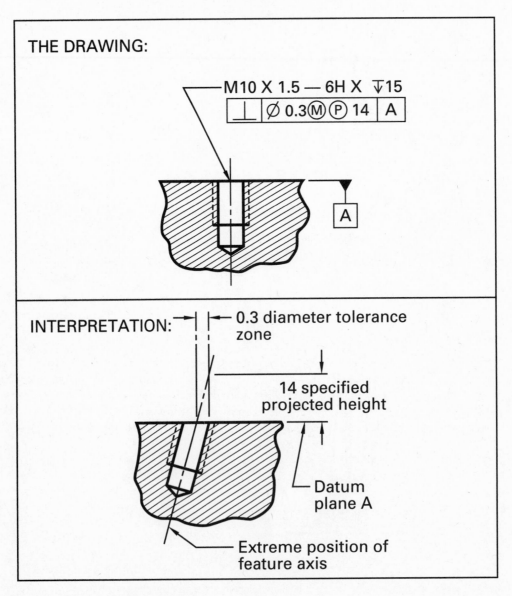

Fig. 18-5   Application of projected tolerance zone.

When the fasteners are studs or press-fit pins, the projected tolerance zone must be equal, not to the thickness of the mating part, but to the installed height of the stud or pin above the surface (see Fig. 18-6).

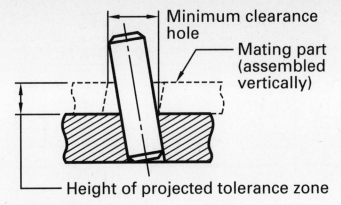

Fig. 18-6   Height of projected tolerance zone.

18-5   ZERO POSITION TOLERANCE AT MMC

Sometimes a part may be rejected because a feature exceeds the MMC size limit, even though it might actually fit the mating part. An example is shown in Fig. 18-7. The minimum size (MMC) is given as ∅14.25 mm to clear a 14 mm screw. If the produced size were ∅14.10 mm (0.10 mm over the maximum screw size) and the actual position error were under 0.10 mm, the screw would fit in the hole. However, the part, although usable, would have to be rejected because of the undersized hole. The unused position tolerance cannot be added to the size tolerance.

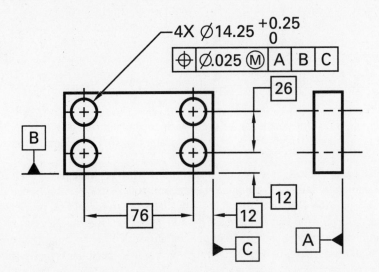

Fig. 18-7   A normal application of position tolerance at MMC.

A tolerancing technique has been developed to prevent this condition. It consists of specifying zero position tolerance and expanding the size tolerance on the hole. The minimum hole limit is made the same as or slightly larger than the maximum mating part. See Fig. 18-8, where the same part in Fig. 18-7 is shown with a different tolerance for the holes. Now all the tolerance, which has been increased to 0.5 mm, is in the hole size. The position tolerance is zero, but that only applies when the holes, as produced, are minimum (MMC), which is unlikely. The actual position tolerance on the produced part will be equal to the amount the actual hole size exceeds the MMC, which can be 0.5 mm, a practical amount. There will be no unused position tolerance. The table on page 102 shows the position tolerance available for various produced hole sizes.

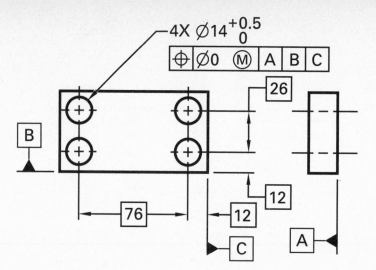

Fig. 18-8   Application of zero tolerance at MMC—same part.

Produced Hole Size (mm)	Clearance with Screw (mm)	Allowable Position Tolerance (mm)
∅14.0	0	0
∅14.1	0.1	0.1
∅14.2	0.2	0.2
∅14.3	0.3	0.3
∅14.4	0.4	0.4
∅14.5	0.5	0.5

A slightly larger drill or punch can be used by the manufacturing department (but within tolerance) to produce the holes in Fig. 18-8 to ensure that some position tolerance will always be available.

In many applications, zero position tolerance at MMC is unsuitable, such as when a specific running or sliding fit is required between mating parts. An often-stated disadvantage of zero position tolerance is that, if both mating parts should be produced at their MMC sizes, the position and form would have to be perfect. But this is a remote possibility. Sometimes the zero tolerance might give an unjustified impression of superior precision and high cost. This problem is generally overcome as personnel receive training and practical experience in zero position tolerancing.

18-6   NONPARALLEL HOLES

Holes that are not parallel and not drilled or punched perpendicular to a flat surface can also be controlled by a position tolerance. An example is shown in Fig. 18-9. Here, radial holes have been drilled in the wall of a tubular part. Each of the holes must be within a .010-diameter tolerance zone. The tolerance zones for each of the features are located 1.000 inch from datum B and centered on 90° axes generated by simulating datum A.

THE DRAWING:

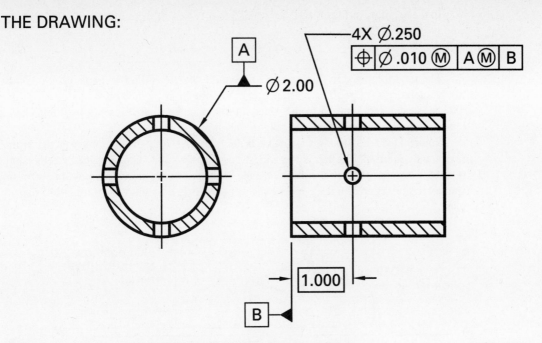

Fig. 18-9   An example of nonparallel holes controlled by position tolerance.

18-7   NONCIRCULAR FEATURES

The benefits of position tolerancing for circular features apply equally well to noncircular features such as slots and tabs. Features of this kind are located not on an axis but on a center plane. Figure 18-10 shows mating parts with slots and tabs controlled by position tolerances. The tolerances were determined by the following formula:

$$T_{tot} = W_s - W_t$$

Where: T_{tot} = Total tolerance in both parts
W_s = Width of slot at MMC
W_t = Width of tab at MMC

Because $W_s - W_t$ equals the minimum clearance between the slot and tab, the total tolerance equals the minimum clearance. This is similar to the floating fastener situation (Section 18.2), except that only two parts are involved (no fasteners).

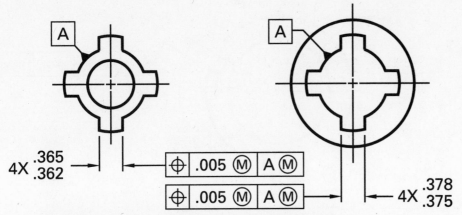

Note: The position tolerance in both drawings is controlling position of the slots or tabs relative to the axis of datum A.

Fig. 18-10   Mating parts with noncircular features controlled by position tolerance.

In the example in Fig. 18-10, W_s = .375 and W_t = .365.

$$T_{tot} = .375 - .365 = .010$$

The total tolerance in this case is split equally between the two parts, each one having a tolerance of .005. The tolerance can be split in any other combination (.007 and .003, .006 and .004, and so forth). Unequal tolerance combinations can be done when a close tolerance is more difficult to hold in one of the parts.

Figure 18-11 illustrates a typical tab and slot of each part assembled in the virtual (extreme) condition. The tab is the widest possible size and slanted at the extreme angle permitted by the geometric tolerance zone. The slot is its narrowest possible size and slanted to the extreme in the opposite direction. This is the tightest possible fit of the two parts.

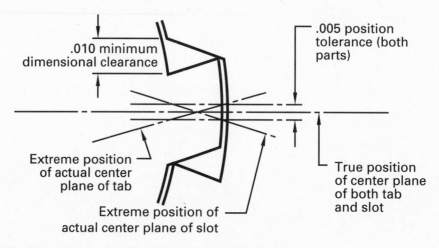

Fig. 18-11 Tightest fit of slot and tab shown in Fig. 18-10.

18-8 USE OF POSITION TOLERANCE FOR ANGULARITY AT MMC

Angularity can be specified MMC when size features are involved. However, this is done more conveniently by controlling the feature and/or the datum by position tolerance. An example is shown in Fig. 18-12. Here both datums and the controlled keyway apply at MMC.

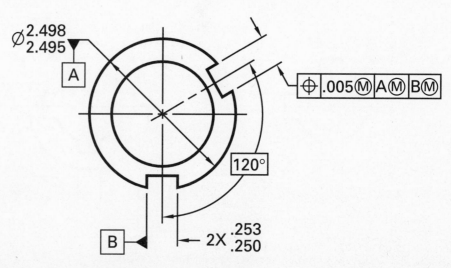

Fig. 18-12 Use of position tolerance to control angularity.

18-9   USE OF POSITION FOR SYMMETRY

Symmetry is the quality of being the same on both sides in size, shape, and relative position from a center plane. The human body is symmetrical on the left and right sides of an imaginary vertical center plane (not front and rear, not top and bottom).

Symmetry applies only to size features. The datum must also be a size feature. *Symmetry error* can be thought of as the amount by which opposite sides of a size feature are unequally spaced from the center plane of the datum. Symmetry tolerance, then, is the total permissible error in symmetry.

The form of the tolerance zone is the same as for any feature located on a center plane, with a slight difference. It is the space between two imaginary planes perfectly parallel to and equidistant from the true position of the center plane *of the datum* (see Fig. 17-2).

As usual, the requirements of the design dictate how the features are controlled. However, when using position to control symmetry, MMC is usually applied to the controlled feature. This situation most often represents a mating condition. The datum can be applied regardless of feature size or at MMC. The controlled feature or the datum is rarely specified at LMC. Figures 18-13 and 18-14 are examples of the use of position to control symmetrical relationships.

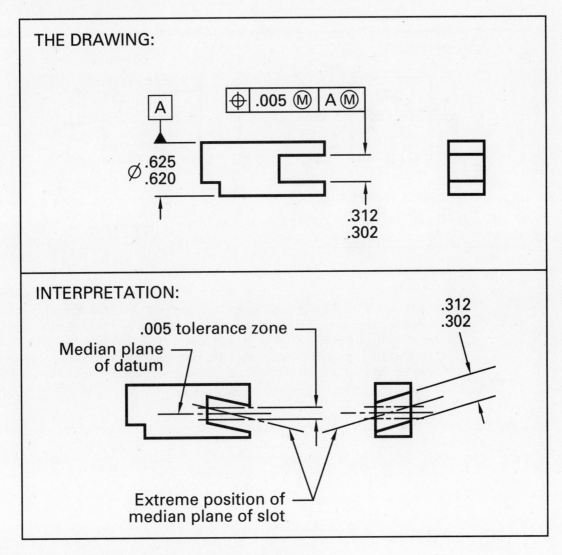

Fig. 18-13   Use of position tolerance to control symmetry of a slot.

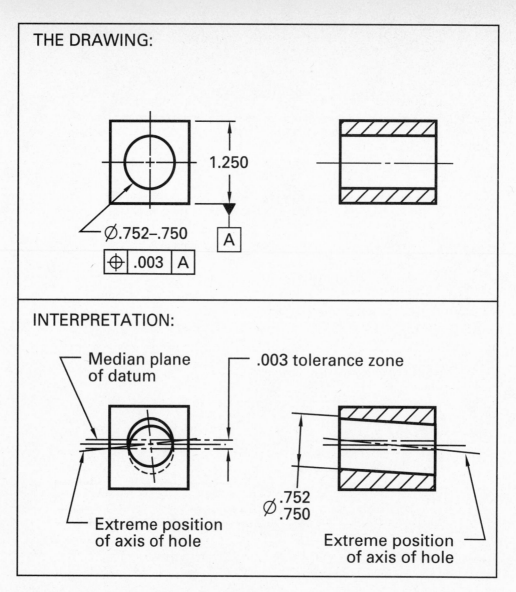

Fig. 18-14   Use of position tolerance to control symmetry of a hole.

When not using a coordinate measuring machine to measure symmetry, it is difficult to measure from a center plane. Therefore, symmetry can be measured to the edge of the feature, as shown in Fig. 18-15. A measurement is taken from one side of the feature to one side of the datum. The part is rotated 180° and another measurement is taken from the other side of the feature to the other side of the datum. The difference between the two measurements is equal to the error in symmetry.

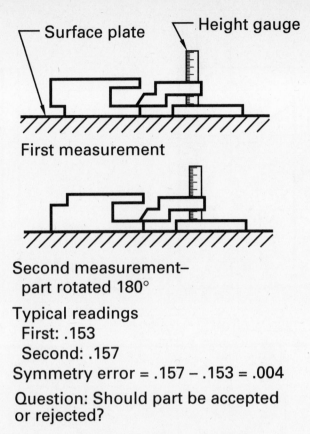

First measurement

Second measurement–
part rotated 180°

Typical readings
  First: .153
  Second: .157
Symmetry error = .157 − .153 = .004

Question: Should part be accepted
or rejected?

Fig. 18-15   Typical symmetry inspection of part shown in Fig. 18-13

Zero position tolerance at MMC can be applied to symmetry as easily as it is to other types of position tolerance. Taking the part in Fig. 18-13 as an example, the position tolerance could be reduced to .000 at MMC. The actual permissible position error would then be equal to the amount that the feature and datum sizes, as produced, depart from MMC. The maximum would be the sum of the size tolerances (.005 + .010 = .015). In the unlikely event that the datum was produced .625 and the slot was produced .302 (both at MMC), the symmetry would have to be perfect.

18-10   MULTIPLE PATTERNS OF FEATURES

Patterns of features are any arrangement of features on an object. The features are usually circular holes, but they can be any kind of opening (round-end or rectangular slots, or any irregular shape), or they can be protrusions such as bosses, pads, or embossed figures, or they can be simply markings on the surface. Although all the following examples will show *hole* patterns, the information applies equally well to all types of feature patterns. When there are two or more feature patterns on the same part, they are called *multiple patterns of features.*

Multiple patterns of features located by basic dimensions from the same datums (but *not* size datums) are considered one composite pattern. (See Figure 18-16.) The four-hole rectangular pattern and the six-hole circular pattern may be inspected with one gage, as if they were one ten-hole pattern. If, because of the complexity of the part or its size, it is necessary to use two gages, the two hole patterns are still treated as one pattern. Both gages must seat against datums A, B, and C.

THE DRAWING:

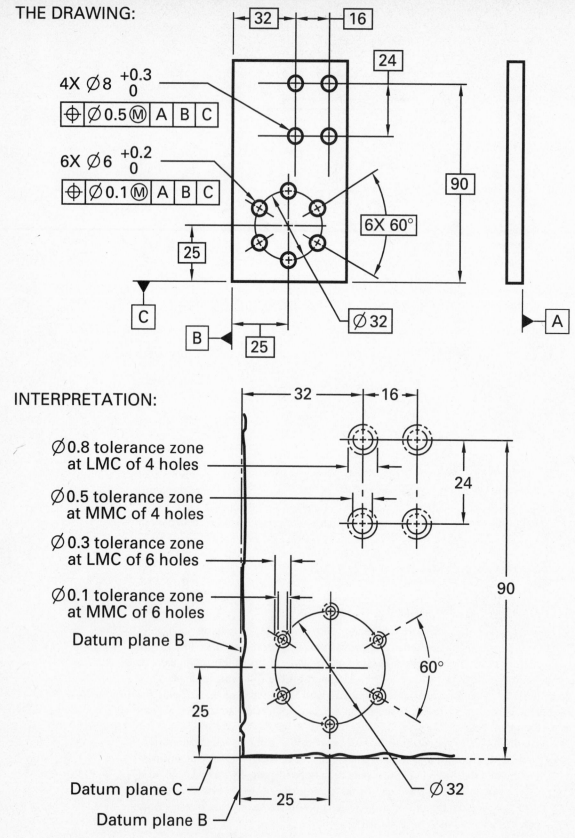

INTERPRETATION:

Ø0.8 tolerance zone at LMC of 4 holes

Ø0.5 tolerance zone at MMC of 4 holes

Ø0.3 tolerance zone at LMC of 6 holes

Ø0.1 tolerance zone at MMC of 6 holes

Datum plane B

Datum plane C

Datum plane B

Fig. 18-16   A four-hole pattern and a six-hole pattern treated as one.

Multiple patterns of features located by basic dimensions from the same datums that *are* size datums are also considered one pattern when their respective feature control frames are identical except for the tolerances. The tolerances can be the same or different. The datums must be specified in the same order and the modifiers must agree for both or all patterns. In Fig. 18-17, the position tolerances for the two 2-hole patterns are different, but everything else in the feature control frames is identical; therefore, the two 2-hole patterns are treated as one pattern.

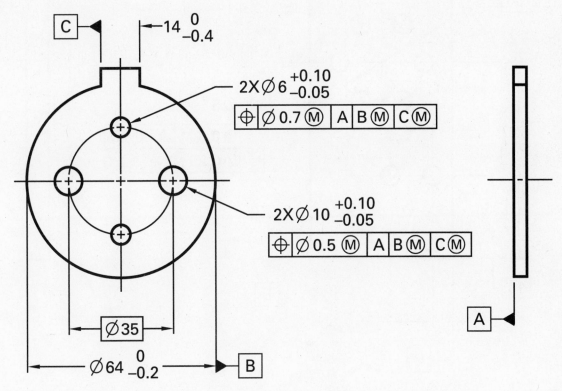

Fig. 18-17   A part with two two-hole patterns located from the same datums, two of which are size features.

In both situations above, if it is required for any reason that the hole patterns be treated as separate requirements and gaged separately, a note "SEP REQT" (separate requirement) is drawn below each feature control frame. This might be done, for example, if the two small holes in Fig. 18-17 are used for attaching another part and the two large holes are used for mounting onto the next assembly.

18-11   COMPOSITE TOLERANCES FOR FEATURE PATTERNS AND INDIVIDUAL FEATURES WITHIN THE PATTERNS

The position location tolerance for a pattern of holes or other features should be larger than the position tolerance of the individual holes. In this case, two feature control frames are drawn as in Fig. 18-18, one below the other, sharing one position tolerance symbol. The upper frame always gives the tolerance for the location of the pattern, and the lower one for the location of individual holes within the pattern.

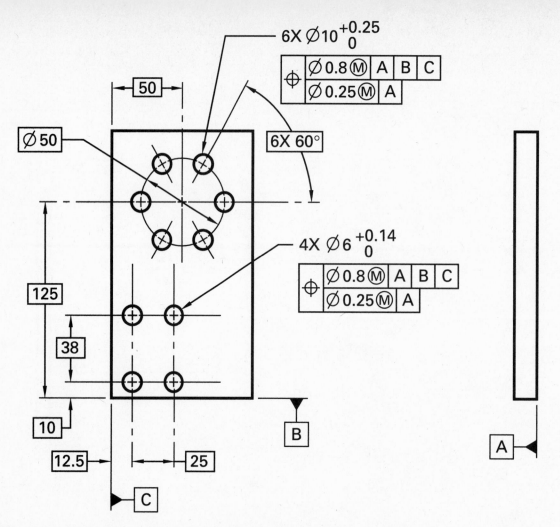

Fig. 18-18   A plate containing two hole patterns and composite position tolerances.

In Fig. 18-18, three datums are used for the pattern location, of which A is the first datum (the most important) because the part will be seated on that surface in the next assembly. Datums B and C are the second and third datums, respectively. They locate the hole patterns dimensionally and provide the required orientation on the part, both within a ∅0.8 mm tolerance zone. The individual holes are located within each pattern by basic dimensions. Datum A is specified to control perpendicularity within the ∅0.25 mm tolerance zone. An interpretation is given in Fig. 18-19 for the circular pattern of holes. The permissible position variation for the pattern consists of six ∅0.8 mm tolerance zones perfectly centered in their true position. The center of each ∅0.25 mm tolerance zone for the individual holes must be within the ∅0.8 mm tolerance zone and located on the intersection of six equally spaced radial lines and a ∅50 mm circle. The actual axis of each hole must be somewhere within the ∅0.25 tolerance zone and also within the larger tolerance zone. Note in Fig. 18-19 that the radial lines and circle on which the smaller tolerance zones are centered may be slightly removed from the desired location (shown on the drawing) on which the larger tolerance zones are centered.

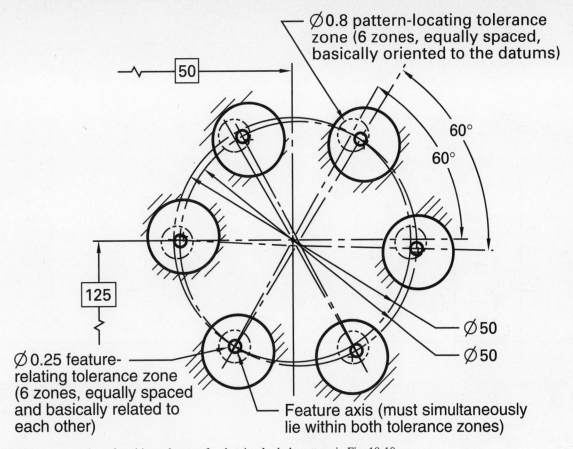

Fig. 18-19   Interpretation of position tolerance for the circular hole pattern in Fig. 18-18.

The interpretation is shown with the holes at their MMC size (smallest diameter). When the holes are at their LMC size, the position tolerance of the pattern may increase to 1.05 diameter and the tolerances of the individual holes may increase to 0.50 diameter.

CHAPTER 19

Position-Coaxial Applications

19-1 DEFINITION

Two or more geometric figures are said to be coaxial if they have the same axis or if their axes are in line (Section 15.1). Any two *regular* shapes can be coaxial. This includes figures such as cylinders, cones, curved profiles, and square or hexagonal shapes, all of which are illustrated in Fig. 15-1. This chapter will be based mostly on cylinders, but all of the information applies equally well to other regular shapes.

When discussing coaxiality in Chapters 15 and 16, we said that there is no symbol for the general case of coaxiality. But there are three symbols that define special cases: concentricity, runout, and position tolerance.

19-2 FORM OF THE TOLERANCE ZONE

The form of the tolerance zone for coaxial position tolerance is an imaginary cylinder perfectly coaxial with the datum axis. The diameter of the cylinder is equal to the specified tolerance; its length is equal to the length of the feature (see Fig. 19-1). This is the same form as for any feature located by its axis for any type of geometric tolerance. The tolerance is always specified as a diameter ($\emptyset$) in the feature control frame.

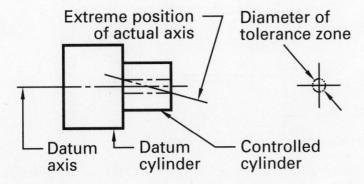

Fig. 19-1   The form of the tolerance zone for coaxial position tolerance.

19-3 SURFACE AND ORIENTATION ERRORS

Coaxial position tolerance is a composite tolerance. In addition to coaxiality, it includes errors on the surface of the feature, such as circularity and straightness. Runout is also a composite tolerance, but it is different from concentricity, which is pure coaxiality.

19-4   USE OF MODIFIERS

Position is the most common method for controlling coaxiality for most applications. It is easier to inspect than concentricity. The tolerance is most often specified at MMC. Runout, which also controls coaxiality, is always applied RFS. A comparison of position tolerance, runout, and concentricity is made at the end of this chapter.

MMC is specified for the feature and for a size datum whenever possible so that additional tolerance is permitted when the size as produced is not at MMC. When the reason for specifying coaxiality is "assemblability" (so that parts will assemble properly), the MMC modifier is appropriate. The "circle M" symbol, Ⓜ, is drawn in the feature control frame after the tolerance and, when desired, after the datum letter. An example is shown in Fig. 19-2.

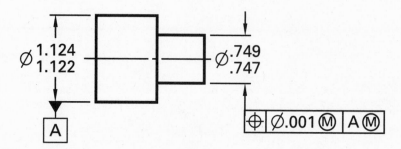

Fig. 19-2   Coaxial position tolerance at MMC.

LMC is not often specified in coaxiality applications. In situations where a wall thickness is more critical than coaxiality, LMC can be used to maintain a certain minimum thickness. In the part shown in Fig. 19-3, the minimum wall thickness, considering the size tolerances and the position tolerance at LMC, is 1.85 mm.

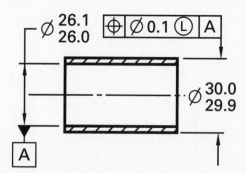

Fig. 19-3   Coaxial position tolerance at LMC.

As the produced sizes of the outside diameter (O.D.) and inside diameter (I.D.) depart from LMC, more position tolerance is permitted, which is shown for three possible sizes in the table below.

Produced Size (mm)		Total Departure from LMC (mm)	Permissible Position Error (mm)	Minimum Wall Thickness (mm)
I.D.	O.D.			
26.1	29.9	0	0.1	1.85
26.05	29.95	0.1	0.2	1.85
26.0	30.0	0.2	0.3	1.85

The total departure from LMC for both the I.D. and O.D. is 0.2 mm, which can be added to the position tolerance as a *bonus tolerance* with no decrease in the critical wall thickness.

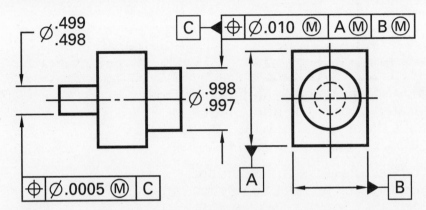

Fig. 19-4 Coaxiality between a rectangular block and two cylinders—MMC and RFS specified.

19-5 SPECIFYING COAXIAL POSITION TOLERANCE

Coaxial position can be applied regardless of feature size by not applying a modifier to the tolerance and/or the datum letter(s). In Fig. 19-4, the position tolerance on the smaller cylinder applies at MMC but datum C is RFS.

Examples of drawings in which coaxiality is controlled by position tolerance are given in Figs. 19-2 through 19-7. Fig. 19-4 shows a cylinder (C) coaxial with a rectangular block in the middle of the part. Both the cylinder and the block apply at MMC. The smaller cylinder on the left is coaxial with cylinder C. Note that, in the feature control frame for the smaller cylinder, the position tolerance applies at the MMC (∅.499) of the smaller cylinder but regardless of the size of datum C, the larger cylinder.

19-6 COAXIAL HOLES

The alignment of two or more holes on a common axis can be controlled by coaxial position tolerance. It is used when a location tolerance alone does not provide the necessary control of coaxiality. Figure 19-5, pp. 115-116, shows an example of four coaxial holes of the same diameter. The pattern of holes must fit within a 0.25 diameter tolerance zone relative to datums A and B. The hole-to-hole relationship must also be within a 0.15-diameter tolerance zone.

THE DRAWING:

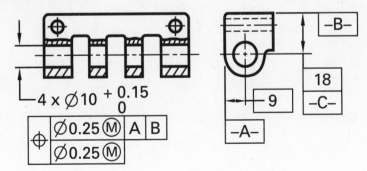

Fig. 19-5 Control of coaxial holes by position.

INTERPRETATION:

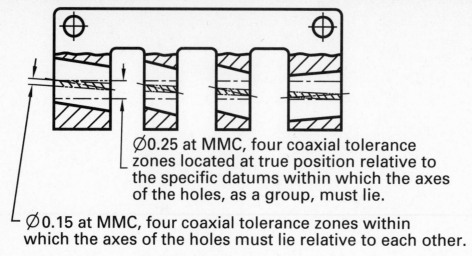

$\emptyset$0.25 at MMC, four coaxial tolerance
zones located at true position relative to
the specific datums within which the axes
of the holes, as a group, must lie.

$\emptyset$0.15 at MMC, four coaxial tolerance zones within
which the axes of the holes must lie relative to each other.

Fig. 19-5   Control of coaxial holes by position *(continued)*.

19-7   CALCULATIONS TO DETERMINE COAXIAL TOLERANCE

To determine the correct position tolerances to use with mating coaxial parts, the designer first selects the dimensional tolerances for each diameter of both parts. This will result in certain maximum and minimum clearances between the parts. To avoid interference, the total position tolerance for both parts must not be greater than the minimum clearance.

Figure 19-6 will be used as an example. Drawings of two mating coaxial parts are shown. The plug is to fit into the socket at both diameters with a metal-to-metal fit at the tightest condition. The loosest condition will be determined by adding the differences of the MMC of the mating features. In part (a) of the illustration, the drawings are shown with dimensional size tolerances already assigned but with no position tolerances. The following three-step procedure will be used to obtain appropriate coaxial position tolerances for both parts (in the general case the opening in one of the parts is referred to as "the hole," and the part that fills the opening is called "the shaft"):

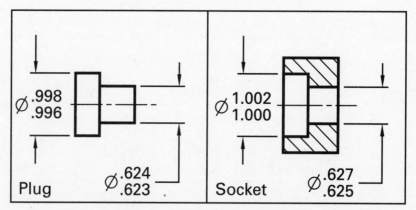

(a) Drawings before positional tolerances are assigned.

Fig. 19-6   Mating plug and socket with calculated position tolerances to ensure desired fit. *(continues)*

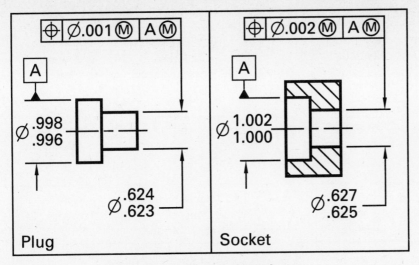

(b) Completed drawings.

Fig. 19-6   Mating plug and socket with calculated position tolerances to ensure desired fit. *(continued)*

1. Obtain the minimum clearance between each set of mating diameters on the two parts.

Socket, minimum hole	.625	1.000
Plug, maximum shaft	−.624	−.998
Minimum clearance	.001	.002

2. Add the minimum clearances. The sum is the total position error that can be allowed for *both* parts.

   $$.001 + .002 = .003$$

3. Divide the total position error between the two parts, not necessarily into equal shares. Of the .003 total tolerance, assign .001 to the shaft (plug) and .002 to the hole (socket). More tolerance is generally allowed for the hole because coaxiality is more difficult to maintain on internal operations, due to fixed tooling sizes.

Although the minimum clearance between the two parts is .001 (at the small diameters) the plug and socket will actually make line contact when they are both at their MMC (largest shaft; smallest hole) because of the out-of-coaxiality allowed by the position tolerances.

   Part (b) of Fig. 19-6 shows the completed drawing with position tolerances added. The smaller diameter is the controlled feature in both drawings and the larger diameter is the datum, but this could have been reversed with no effect on the fit. Both the controlled features and the datums apply at their MMC, so as the produced sizes depart from MMC, more position tolerance is allowed. The tables on page 118 show the effect of certain produced sizes on the position tolerances. Not all possible combinations of controlled feature size and datum size are shown.

Plug

Controlled Feature	Datum	Total Departure from MMC (Bonus Tolerance)	Permissible Position Error
∅.624	∅.998	0	∅.001
∅.6235	∅.997	.0015	∅.0025
∅.623	∅.996	.003	∅.004

Socket

Controlled Feature	Datum	Total Departure from MMC (Bonus Tolerance)	Permissible Position Error
∅.625	∅1.000	0	∅.002
∅.626	∅1.001	0.002	∅.004
∅.627	∅1.002	.004	∅.006

Although the drawings specify position tolerances of .001 and .002, the inspector can accept as much as .004 and .006, respectively, without sacrificing proper fit. Because most parts are produced somewhere between the high and low size limits, the probable position tolerances permissible will be between the extremes—.002 for the plug and .003 for the socket.

19-8   ZERO TOLERANCE AT MMC

Zero tolerance at MMC can be used as effectively with coaxial features as with other types of position situations. Figure 19-7 is a drawing of the same part as in Fig. 19-2. Here the small cylinder must be perfectly coaxial (zero position tolerance) with the datum cylinder when both features are at MMC. The size tolerances on both features, however, allow a total variation of .005, and this can be used as position tolerance. If both cylinders were produced at their *least* material condition (∅.747 and ∅1.122), the coaxiality error would be a maximum of ∅.005. The possibility that both features will be produced at their MMC sizes is remote. In all likelihood they will each be about .001 under the maximum limit, providing a coaxiality tolerance of about ∅.002. As written in Section 18.5, zero position tolerancing is a practical alternative method of controlling position variations while avoiding unused tolerances and possible rejection of usable parts.

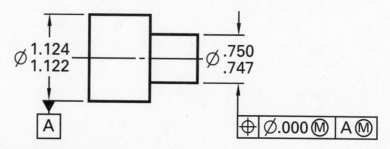

Fig. 19-7   Specifying zero tolerance at MMC.

19-9   SELECTION OF PROPER CONTROL FOR COAXIALITY

Use position tolerance when the following conditions apply:

1. Surface errors may be included with coaxiality error.

2. The coaxial features must assemble with another part having corresponding coaxial features, and additional coaxiality tolerance can be allowed when the feature and/or the datum are not at the maximum material condition (see Section 19.4).

Use concentricity when the following conditions apply:

1. Coaxiality must be controlled independently of surface errors.
2. The desired coaxiality tolerance must be held regardless of feature and datum size (RFS).

Use runout when the following conditions apply:

1. Surface errors may be included with coaxiality error.
2. The desired coaxiality tolerance must be held regardless of feature and datum size (RFS).

CHAPTER 20

Symmetry

20-1 DEFINITION

Symmetry is the condition when the center plane of a rectangular feature (slot or tab) is congruent with the center plane of a rectangular size datum (see Fig. 20-1). The datum is usually on the outside of the controlled feature. The center plane is composed of all the median points measured between the opposing surfaces of the controlled feature.

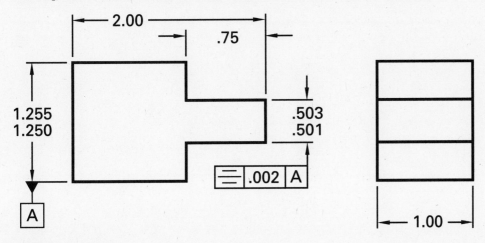

Fig. 20-1 Symmetry.

20-2 SYMMETRY TOLERANCE ZONE

The tolerance zone for symmetry is the space between two parallel planes equally disposed about the center plane of the datum. Each of the median points measured between the opposing surfaces of the controlled feature must be within the parallel planes.

20-3 SYMMETRY USE

Symmetry can be thought of as a special case of position. It can be applied only on an RFS basis. It is usually applied to create an equal distribution of mass for purposes of wall thickness, strength, balance, etc., without the subsequent adding or subtracting of material to achieve the equilibrium. Because of the potentially complex and time-consuming nature of inspection, symmetry is not commonly used.

Appendixes

APPENDIX A

Summary of Nonstandard Practices

A-1 INTRODUCTION

Many thousands of drawings still treat geometric dimensioning differently from the methods specified in ASME Y14.5M-1994. Many design groups are not yet using the current ASME standard. For these reasons, this appendix has been added to explain nonstandard practices that are still applied to current drawings and that are found on older drawings still in use for production, rework, or spare parts. The intent of these geometric specifications can be explained in the title block, the general notes, the company drafting room manual (DRM), or in a manufacturing standard. Sometimes the intent may have been only in the mind of the designer.

A-2 FORMERLY USED SYMBOLS

The symbols for straightness and flatness have undergone some evolution, as shown in Fig. A-1. The older symbols were used in MIL-STD-8, which was the standard on dimensioning and tolerancing used by United States government agencies until 1966, when the government adopted USASI Y14.5.

Characteristic	Former Symbol	Present Symbol
Straightness	⌒	—
Flatness	～ OR ⌒ OR —	▱
Parallelism	‖	∥
Concentricity	⊙	◎
Circular Runout	[↗] Circular	↗
Total Runout	↗ OR [↗] Total	↗↗
Datum Feature	–A–	A ▲

Fig. A-1 Formerly used symbols.

The concentricity symbol on many drawings was intended to denote coaxiality including surface errors, which is now called runout. When such a drawing is used and runout was actually intended, the drawing should be treated as though the geometric characteristic were runout. If the intent is not known, the concentricity symbol must be taken literally and the part produced and inspected for true concentricity. An optional form of the concentricity symbol had a filled-in inner circle. This had no effect on the meaning.

Considerable confusion surrounds runout. The symbol ↗, which is now standardized as circular runout, *has* been used to denote total runout. If circular runout was in fact intended, the word CIRCULAR was added below the feature control frame. The same symbol in other places was used in the opposite way. In the 1973 edition of ANSI Y14.5, the symbol ↗, unless otherwise specified, denoted circular runout, as it does now. Total runout was specified by adding the note TOTAL below the feature control frame.

A-3   FEATURE CONTROL FRAME

The feature control frame was formerly called the feature control *symbol* and has passed through other phases before acquiring its present configuration. Originally, the frame had no partitions and the sequence of entries was different—the datum letter and the tolerance were reversed. The abbreviation DIA was used in place of the symbol ∅ [see Fig. A-2 (a)]. Later, partitions were added but the sequence of entries remained the same (reversed from present use), as shown in Fig. A-2 (b).

(a)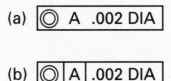

(b)

Fig. A-2   Former configurations of the feature control frame.

Tolerances for geometric characteristics that can be inspected by the use of a dial indicator were once labeled TIR (total indicator reading) or FIR (full indicator reading) (see Fig. A-3). The abbreviations TIR and FIR were identical in meaning to the current standard FIM (full indicator movement).

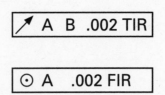

Fig. A-3   Former designation of diametral tolerance.

A-4   SIZE OF THE TOLERANCE ZONE

The concept of the size of the tolerance zone has varied. In modern practice, the tolerance given in the feature control frame is the *total* width or the *diameter* of the tolerance zone. Some designers in the past thought of the permissible error as the deviation from perfect geometry in *either* direction, so the tolerance given was *half* the total width for a rectangular tolerance zone. Sometimes "R" was specified with a cylindrical tolerance, but there was no label indicating that a rectangular tolerance was half the total width. When a tolerance was meant to apply to the total width or the diameter, the word TOTAL or the abbreviation DIA was added after the tolerance value. TOTAL was often abbreviated TOT.

A-5   LOCATION OF FEATURE PATTERNS

On many existing drawings, feature patterns are located by dimensions with plus-and-minus tolerances from unlabeled (implied) datums, while the individual features are controlled by basic dimensions and position tolerances. The standard method is to use basic dimensions and position tolerances for both the pattern location and individual feature location, as explained in Chapter 18. An example of hole patterns located by dimensions with plus-and-minus tolerances is shown in Fig. A-4, and the interpretation of the tolerance zones for the circular pattern of holes is given in Fig. A-5. This part is similar to that in Fig. 18-20 and Fig. 18-21. The fault with toleranced coordinate dimensions is that they produce a square tolerance zone wherein the deviation from the true position (the center of the square) is not the same in all directions (see Chapter 17).

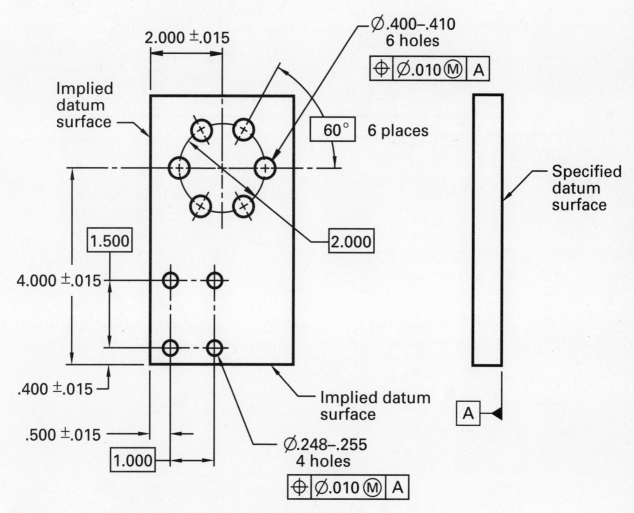

Fig. A-4   A plate with hole patterns located by toleranced dimensions.

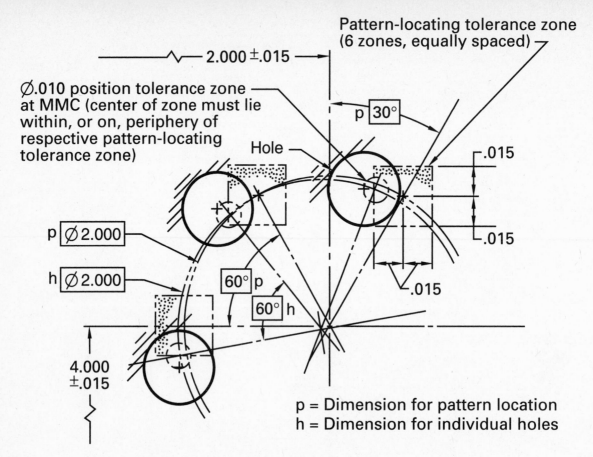

Fig. A-5   Interpretation of the tolerance zones for the circular hole pattern in Fig. A-4.

APPENDIX B

Conversion Tables— Position Tolerance to Coordinate Tolerance

B-1 INTRODUCTION

Quality assurance personnel must often convert coordinate tolerances to position tolerances. Table B-1 is provided for this purpose.

B-2 CONVERSION OF COORDINATE MEASUREMENTS TO POSITION TOLERANCE

Figure B-1 shows a typical situation for one feature located by basic dimensions from two datums. The location of the feature is inspected by a coordinate measurement machine from the two datums on the actual part, and the difference between the actual measurement and the basic dimension in each direction is noted.

POSITIONAL TOLERANCE DIAMETER Z

Y \ X	.001	.002	.003	.004	.005	.006	.007	.008	.009	.010	.011	.012	.013	.014	.015	.016	.017	.018	.019	.020
.020	.0400	.0402	.0404	.0408	.0412	.0418	.0424	.0431	.0439	.0447	.0456	.0466	.0477	.0488	.0500	.0512	.0525	.0538	.0552	.0566
.019	.0380	.0382	.0385	.0388	.0393	.0398	.0405	.0412	.0420	.0429	.0439	.0449	.0460	.0472	.0484	.0497	.0510	.0523	.0537	.0552
.018	.0360	.0362	.0365	.0369	.0374	.0379	.0386	.0394	.0403	.0412	.0422	.0433	.0444	.0456	.0469	.0482	.0495	.0509	.0523	.0538
.017	.0340	.0342	.0345	.0349	.0354	.0360	.0368	.0376	.0385	.0394	.0405	.0416	.0428	.0440	.0453	.0467	.0481	.0495	.0510	.0525
.016	.0321	.0322	.0325	.0330	.0335	.0342	.0349	.0358	.0367	.0377	.0388	.0400	.0412	.0425	.0439	.0452	.0467	.0482	.0497	.0512
.015	.0301	.0303	.0306	.0310	.0316	.0323	.0331	.0340	.0350	.0360	.0372	.0384	.0397	.0410	.0424	.0439	.0453	.0469	.0484	.0500
.014	.0281	.0283	.0286	.0291	.0297	.0305	.0313	.0322	.0333	.0344	.0356	.0369	.0382	.0396	.0410	.0425	.0440	.0456	.0472	.0488
.013	.0261	.0263	.0267	.0272	.0278	.0286	.0295	.0305	.0316	.0328	.0340	.0354	.0368	.0382	.0397	.0412	.0428	.0444	.0460	.0477
.012	.0241	.0243	.0247	.0253	.0260	.0268	.0278	.0288	.0300	.0312	.0325	.0339	.0354	.0369	.0384	.0400	.0416	.0433	.0449	.0466
.011	.0221	.0224	.0228	.0234	.0242	.0250	.0261	.0272	.0284	.0297	.0311	.0325	.0340	.0356	.0372	.0388	.0405	.0422	.0439	.0456
.010	.0201	.0204	.0209	.0215	.0224	.0233	.0244	.0256	.0269	.0283	.0297	.0312	.0328	.0344	.0360	.0377	.0394	.0412	.0429	.0447
.009	.0181	.0184	.0190	.0197	.0206	.0216	.0228	.0241	.0254	.0269	.0284	.0300	.0316	.0333	.0350	.0367	.0385	.0402	.0420	.0439
.008	.0161	.0165	.0171	.0179	.0189	.0200	.0213	.0226	.0241	.0256	.0272	.0288	.0305	.0322	.0340	.0358	.0376	.0394	.0412	.0431
.007	.0141	.0146	.0152	.0161	.0172	.0184	.0198	.0213	.0228	.0244	.0261	.0278	.0295	.0313	.0331	.0349	.0368	.0386	.0405	.0424
.006	.0122	.0126	.0134	.0144	.0156	.0170	.0184	.0200	.0216	.0233	.0250	.0268	.0286	.0305	.0323	.0342	.0360	.0379	.0398	.0418
.005	.0102	.0108	.0117	.0128	.0141	.0156	.0172	.0189	.0206	.0224	.0242	.0260	.0278	.0297	.0316	.0335	.0354	.0374	.0393	.0412
.004	.0082	.0089	.0100	.0113	.0128	.0144	.0161	.0179	.0197	.0215	.0234	.0253	.0272	.0291	.0310	.0330	.0349	.0369	.0388	.0408
.003	.0063	.0072	.0085	.0100	.0117	.0134	.0152	.0171	.0190	.0209	.0228	.0247	.0267	.0286	.0306	.0325	.0345	.0365	.0385	.0404
.002	.0045	.0056	.0072	.0089	.0108	.0126	.0146	.0165	.0184	.0204	.0224	.0243	.0263	.0283	.0303	.0322	.0342	.0362	.0382	.0402
.001	.0028	.0045	.0063	.0082	.0102	.0122	.0141	.0161	.0181	.0201	.0221	.0241	.0261	.0281	.0301	.0321	.0340	.0360	.0380	.0400

Y ↕   X ↔

Table B-1 Conversion of coordinate measurements to position tolerance.

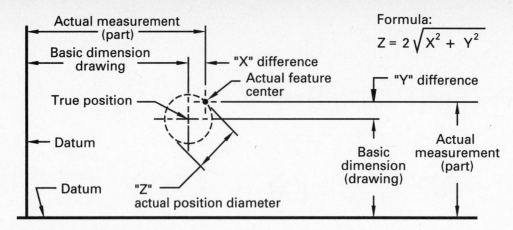

Fig. B-1   Position tolerance calculation.

In the drawing, the differences are labeled X difference and Y difference. The Z diameter is the corresponding position tolerance diameter. All the figures in the body of the table are Z values.

EXAMPLE:

The drawing of a plate requires that a hole of diameter .374–.380 must be in true position relative to two datums within $\varnothing$.014 at MMC. The location of the hole is measured, and the axis is found to be .005 from true position in the X direction and .006 off in the Y direction. Should the part be accepted or rejected?

Enter the table at the .005 column on the X scale and move up to the row headed .006 on the Y scale. Find the Z value, .0156. This is the diameter of the equivalent position tolerance zone. If the tolerance were applied RFS, the part would be rejected. However, because the tolerance is specified MMC, a bonus tolerance may make the part acceptable if the hole as produced is larger than its MMC.

The actual diameter of the hole is measured and is found to be .377 diameter, which exceeds the MMC (.374) by .003, the bonus tolerance. Add .003 to the .014-diameter position tolerance given on the print and obtain .017 diameter as the *total* permissible position tolerance. Because the actual hole location is within a .0156-diameter circle, it is inside the .017-diameter limit, and the part is acceptable.

The conversion method described above works equally well for any number of features located by position tolerances. Table B-1 can be used to verify the location of feature patterns such as hole patterns that are specified on drawings with composite position tolerances (see Chapter 18). In this case, two conversions must be made: one for the location of the pattern as a whole and the other for the location of the individual holes within the pattern.

Index

Chapter 1 Review Problems

Write the letter of the best answer in the space at left.

_____ **1.** If geometric tolerances are not specified on a drawing, what controls the geometry?

    (a) dimensions   **(b)** views   **(c)** material   **(d)** notes

_____ **2.** The Y14.5M standard recommends geometric tolerances be specified by the use of _____ .

    (a) local notes   **(b)** general notes
    (c) symbols   **(d)** size tolerances

_____ **3.** If the upper surface of a block is located by a $\pm.01$ size tolerance from a bottom surface that is absolutely flat, the upper surface can be curved by a maximum of _____ .

    (a) .005   **(b)** .01   **(c)** .02   **(d)** more than .02

_____ **4.** When did the British start using geometric tolerancing symbols?

    (a) 1920s   **(b)** 1930s   **(c)** 1940s   **(d)** 1950s

_____ **5.** What is the abbreviation of the name of the organization that currently governs the development of geometric tolerancing practices?

    (a) ASA   **(b)** USASI   **(c)** NASA   **(d)** ANSI

_____ **6.** What is the year that the current Y14.5 standard was made available to the public?

    (a) 1973   **(b)** 1982   **(c)** 1994   **(d)** 1995

Chapter 2 Review Problems

Next to each geometric characteristic listed below, select and sketch the appropriate symbol from the symbols shown.

_____ 1. Straightness   _____ 8. Flatness

_____ 2. Circularity   _____ 9. Cylindricity

_____ 3. Profile of a line _____ 10. Profile of a surface

_____ 4. Parallelism   _____ 11. Perpendicularity

_____ 5. Angularity   _____ 12. Circular runout

_____ 6. Concentricity   _____ 13. Total runout

_____ 7. Position   _____ 14. Symmetry

_____ 15. The symbol for *diameter* is one of those shown below. Sketch the correct symbol in the space at left.

ϕ   ⊖   ⌀   D

Write the letter of the best answer in the space at left.

_____ 16. Which of the following types of lines cannot have a feature control frame attached to it?

  (a) extension   **(b)** leader   **(c)** center   **(d)** dimension

Chapter 3 Review Problems

Part I

Write the letter of the best answer in the space at left.

_____ 1. What is the name of the area of the total permissible error of a geometric tolerance?

     (a) virtual condition   (b) basic dimension
     (c) datum   (d) tolerance zone

_____ 2. What is the maximum material condition of a hole?

     (a) smallest size   (b) largest size
     (c) nominal size   (d) basic size

_____ 3. What is the maximum material condition of a shaft?

     (a) smallest size   (b) largest size
     (c) nominal size   (d) basic size

_____ 4. Which of the following is the symbol for maximum material condition?

     (a) Ⓡ   (b) Ⓕ   (c) Ⓢ   (d) Ⓜ

_____ 5. The virtual condition is the combination of the maximum material condition and the _____ .

     (a) size tolerance   (b) geometric tolerance
     (c) mating size   (d) bonus tolerance

_____ 6. What is the abbreviation of words that refer to the total movement of the probe of a dial indicator?

     (a) FIR   (b) FIM   (c) TIR   (d) TIM

_____ 7. Which of the following is the symbol used to indicate the tolerance zone is above the surface of the part?

     (a) Ⓟ   (b) Ⓜ   (c) Ⓛ   (d) Ⓔ

_____ 8. What is the name of the theoretically exact surface from which geometric dimensions are measured?

     (a) feature   (b) datum
     (c) datum feature   (d) nominal surface

_____ 9. What is the name of a line, real or imaginary, that can be drawn on a surface?

     (a) element   (b) feature   (c) datum   (d) radial line

_____ 10. What is the name of a line that extends toward or away from a center?

     (a) element   (b) center line
     (c) extension line (d) radial line

Chapter 3 Review Problems   Part II

In the space to the right of each statement, write the word or term from the list below that pertains to the statement.

Any line, real or imaginary, that can be drawn on a surface.

1. _____

Any portion of an object, such as a point, axis, plane, or cylindrical surface, or a tab, recess, or groove.

2. _____

Describes a dimension that does not have a tolerance. The tolerance is given elsewhere.

3. _____

The size of a shaft as produced is such that it contains the most material possible.

4. _____

A theoretically exact surface or line from which dimensions or geometric tolerances can be taken.

5. _____

The geometric tolerance specified applies, no matter how big or small the feature can be produced.

6. _____

The physical state of an object that results in the tightest fit with a mating part.

7. _____

The total reading of a dial indicator.

8. _____

Pointing toward or away from a center.

9. _____

The area taken up by the total amount of permissible geometric error.

10. _____

Select answers from the following:

Tolerance zone	Element	Normality
Virtual condition	Radial line	MMC
Datum	Feature	FIM
Basic	Characteristic	RFS
Median	Modifier	LMC

Chapter 4 Review Problems

Write the letter of the best answer in the space at left.

_____   **1.** Which of the following would be a proper datum?

            (a) lathe center   **(b)** bearing surface
            (c) center line   **(d)** axis

_____   **2.** Which of the following groups of letters are not used by themselves on an engineering drawing?

            (a) I, O, Q   **(b)** A, B, C   **(c)** M, S, P   **(d)** X, Y, Z

_____   **3.** What can be a datum feature?

            (a) feature of a part shown on a drawing
            (b) feature of the tooling used to manufacture the part
            (c) neither (a) nor (b)
            (d) both (a) and (b)

_____   **4.** Where does a datum plane exist?

            (a) on the drawing   **(b)** on the part
            (c) in the tooling   **(d)** in theory only

_____   **5.** What is a simulated datum?

            (a) machine tool mounting surface
            (b) assembly fixture mounting surface
            (c) inspection fixture mounting surface
            (d) any of the above

_____   **6.** Which of the following is not a reason to use a datum reference frame?

            (a) restrict degrees of freedom
            (b) move parts between locations
            (c) repeat location of parts
            (d) establish measurement surfaces

_____   **7.** Datum features should be selected because _____ .

            (a) they are functional   **(b)** they are mating features
            (c) they are readily accessible   **(d)** all of the above

_____   **8.** The MMC size of a hole is when the hole is _____ .

            (a) as large as it can be   **(b)** as small as it can be
            (c) its nominal size   **(d)** none of the above

_____   **9.** Determining the condition of a size datum is primarily based on _____ .

            (a) the relationship between mating parts
            (b) the type of material
            (c) the size of the feature
            (d) whether the feature is internal or external

_____ **10.** A datum target can be a _____ .

    (a) point   **(b)** line   **(c)** surface area   **(d)** any of the above

_____ **11.** Where are the datum target identifying letter and number placed in the datum target symbol?

    (a) in the lower half   **(b)** in the upper half
    (c) above the symbol   **(d)** below the symbol

_____ **12.** What is the name for a datum that consists of two or more individual surfaces used to restrict one principal degree of freedom?

    (a) combined datum   **(b)** multiple datum
    (c) compound datum   **(d)** equalizing datum

_____ **13.** What is the name for a datum used to center a noncircular part?

    (a) combined datum   **(b)** multiple datum
    (c) compound datum   **(d)** equalizing datum

Chapter 4 Review Problems   Part II

1. On the isometric drawing below of an unsupported object in space, select suitable datum targets (tooling points) on the three visible surfaces to make a complete datum frame. Show the point locations with crosses. Omit dimensions.

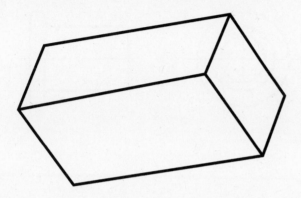

2. Add datum target symbols to both views in the drawing below, complete with datum identifying letters and target numbers.

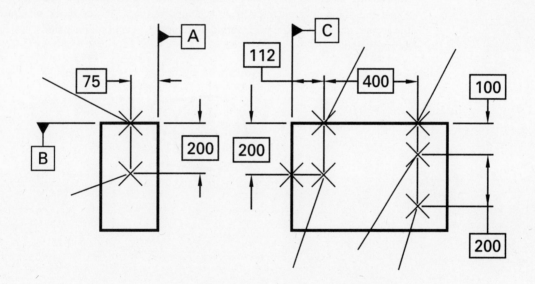

3. The cylindrical part below is mounted between two bearings on the small diameters on each end. Make each small diameter and the left end face separate datums.

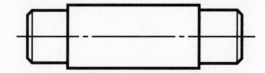

4. Add a complete datum frame to the cast tube below. Select suitable datum targets on cast surfaces. Label the datum planes, give location dimensions for all datum targets, and show all datum target symbols.

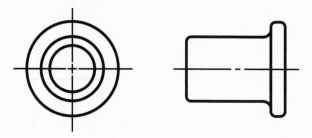

5. In the partial full-size view below, two of three datum targets for the first datum are shown. They are intended to be on the far side (not visible). Add appropriate location dimensions, datum target symbols, and leaders for the two datum targets only. The round end of the object is intended to be supported by a 90° vee-block, which will be the second datum. Add the necessary datum targets, location dimensions, and datum target symbols.

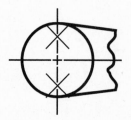

Chapter 5 Review Problems

Write the letter of the best answer in the space at left.

_____ 1. If a dial indicator is used to check parallelism and the FIM is .004, what is the minimum parallelism error?

(a) .002   (b) .004   (c) .005   (d) .008

_____ 2. What is the name of the part on a dial indicator that rotates?

(a) clamp   (b) case   (c) probe   (d) bezel

_____ 3. If the movement observed on a dial indicator is .001 to the left and .003 to the right, what is the FIM?

(a) .001   (b) .003   (c) .004   (d) .008

_____ 4. What does the "F" in FIM stands for?

(a) free   (b) fine   (c) final   (d) full

_____ 5. What lot size would most likely be used for measuring with a dial indicator or coordinate measuring machine?

(a) 10   (b) 1000   (c) 5000   (d) 10,000

_____ 6. A functional gage used for inspecting a machined part simulates the fit of the _____.

(a) mating part   (b) dial indicator
(c) surface plate   (d) vee-block

_____ 7. In what direction does the pointer of a dial indicator move?

(a) clockwise only
(b) counterclockwise only
(c) clockwise and counterclockwise
(d) radially

_____ 8. What is the name of the CMM inspection surface?

(a) surface table   (b) surface plate
(c) measuring surface   (d) none of the above

Chapter 6 Review Problems

Write the letter of the best answer in the space at left.

_____ **1.** What type of tolerance controls the form of a part as well as the size when no geometric tolerance is specified?

    (a) internal   **(b)** external   **(c)** virtual condition   **(d)** size

_____ **2.** When a geometric tolerance is applied to a size feature, which condition applies automatically?

    (a) RFS   **(b)** MMC   **(c)** LMC   **(d)** virtual condition

_____ **3.** When a geometric tolerance is applied to a size feature referenced to a geometrically controlled size datum, which condition applies to the datum feature?

    (a) RFS   **(b)** MMC   **(c)** LMC   **(d)** virtual condition

_____ **4.** Unless otherwise stated, a geometric tolerance applies to which part of a screw thread?

    (a) major diameter   **(b)** minor diameter
    (c) pitch diameter   **(d)** none of the above

_____ **5.** The limits of size rule states that no element of a part shall extend beyond what boundary?

    (a) RFS   **(b)** MMC   **(c)** LMC   **(d)** virtual condition

_____ **6.** The limits of size rule applies only to

    (a) individual features   **(b)** interrelated features
    (c) bar stock   **(d)** twist drills

_____ **7.** The envelope principle makes possible the calculation of a(n) _____.

    (a) allowance   **(b)** maximum clearance
    (c) position tolerance   **(d)** (a) and (b)

Chapter 6 Review Problems   Part II

In the space below, draw the end view of a bar, full size, that is 49.05–50.05 mm wide
(left to right) and 12.00–12.05 mm high. Add limit dimensions.

1. On a bar produced from this drawing, what will be the sizes of the height and width
   at the MMC boundary?

   _____ × _____

2. When the part drawn above is produced at its MMC boundary, what is the permissible
   width and height error in its geometric form?

   _____

3. In the straightness tolerance shown below, no modifier is specified. Which modifier
   applies automatically?

—	.005

   _____

4. In the space below, draw the feature control frame for a screw thread whose pitch
   diameter is to be positioned to datum A within .005 inches. No screw thread drawing
   is necessary.

5. In the space below, draw the feature control frame to control the circular runout of the
   pitch diameter of a gear to datum C within .003 inches. No gear drawing is necessary.

6. In the drawing below, the four holes are positioned relative to datums A and B within .010 inch when the holes and datum A are at MMC. Datum A is a size feature with its own position specification. What is the virtual condition of datum A?

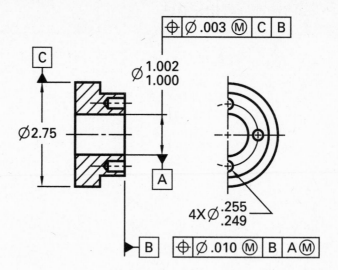

Chapter 7 Review Problems

Part I

Write the letter of the best answer in the space at left.

_____   **1.** What is controlled when straightness is specified?

   (a) surface   **(b)** element   **(c)** chord   **(d)** datum

_____   **2.** What is the configuration of the tolerance zone for straightness applied to a plane feature?

   (a) parallel lines   **(b)** cylindrical   **(c)** square   **(d)** conical

_____   **3.** How is straightness applied unless otherwise specified?

   (a) at MMC   **(b)** at LMC
   (c) RFS   **(d)** to a limited length

_____   **4.** Straightness of a size feature is the only control that will allow the feature to exceed _____ .

   (a) LMC   **(b)** virtual condition
   (c) MMC   **(d)** basic dimension

_____   **5.** What is straightness called when it is applied for a certain distance?

   (a) digital   **(b)** unit   **(c)** dimensional   **(d)** partial

_____   **6.** What is the configuration of the tolerance zone for straightness applied to a cylinder?

   (a) cylinder   **(b)** cone   **(c)** ring
   (d) space between two parallel straight lines

_____   **7.** In which view is straightness of a flat surface applied?

   (a) front view   **(b)** side view   **(c)** where the surface is an area
   (d) where the surface appears as an edge

Chapter 7 Review Problems   Part II

1. In the space below, draw the feature control frame for a straightness tolerance of
   .002 inch applied to a surface.

2. In the space below, draw two views (front and end) of a part that is 2 inches wide
   (left to right), .75 inch high, and .50 inch deep. Omit dimensions. Add a .005-inch element straightness tolerance to the top surface in the front view and a .002-inch element straightness tolerance to the rear surface.

3. In the space below, repeat the front view of the drawing in problem 2 and add a two-
   place height dimension. Show with a properly placed feature control frame that the
   height is to be straight within .003 inch.

4. A long round shaft has a unit size straightness of .001 inch per 1.000-inch length regardless of feature size. Draw the appropriate feature control frame below.

5. A long round shaft has a unit size straightness of .001 inch per 1.000-inch length at MMC, but the total straightness at MMC must be within .010 inch. Draw the appropriate feature control frame below.

Chapter 8 Review Problems

Part I

Write the letter of the best answer in the space at left.

_____ **1.** A flatness tolerance zone is the space between two parallel

_____ .

   (a) planes   **(b)** elements   **(c)** straight lines   **(d)** datums

_____ **2.** To which of the following views of a surface is a flatness feature control frame directed?

   (a) true size   **(b)** horizontal
   (c) vertical   **(d)** an edge view

_____ **3.** What is the measure of how much a plane surface deviates from being a true plane?

   (a) plane deviation   **(b)** plane variation
   (c) flatness error   **(d)** a tolerance zone

_____ **4.** A surface can be straight in one direction but not be flat. The preceding statement _____ .

   (a) requires more data   **(b)** is questionable
   (c) is true   **(d)** is false

_____ **5.** No element of a surface controlled by a flatness tolerance may extend beyond what boundary?

   (a) LMC   **(b)** MMC   **(c)** RFS   **(d)** virtual condition

_____ **6.** What is flatness specified per area called?

   (a) dimensional flatness   **(b)** unit flatness
   (c) partial flatness   **(d)** digital flatness

_____ **7.** What type of line is used to show the boundary of a particular area controlled by flat?

   (a) a break line   **(b)** extension lines
   (c) hidden lines   **(d)** chain lines

Chapter 8 Review Problems

Part II

1. In the space below, draw the feature control frame for a flatness tolerance of .002 inch.

2. In the space below, draw two views (front and end) of a part that is 2 inches wide (left to right), .75 inch high, and .50 inch deep. Omit dimensions. Add a flatness tolerance of .005 inch to the top surface and a flatness tolerance of .002 inch to the rear surface.

3. In the drawing below, with no flatness control specified, what is the maximum permissible flatness error of the top surface if the bottom surface is perfectly flat?

_____

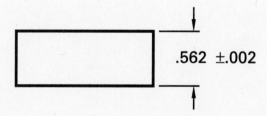

.562 ±.002

4. In the drawing for problem 3, what is the maximum permissible flatness error of the top surface if the bottom surface is out-of-flat by .001 inch?

_____

5. In the space below, redraw the drawing for problem 3 and add flatness tolerances of .001 inch to the top and bottom surfaces.

6. Shown below is one possible enlarged interpretation of the student's drawing for problem 5. Given the size and flatness tolerances of the correct drawing and this interpretation, fill in the values of dimensions A through D in the spaces provided.

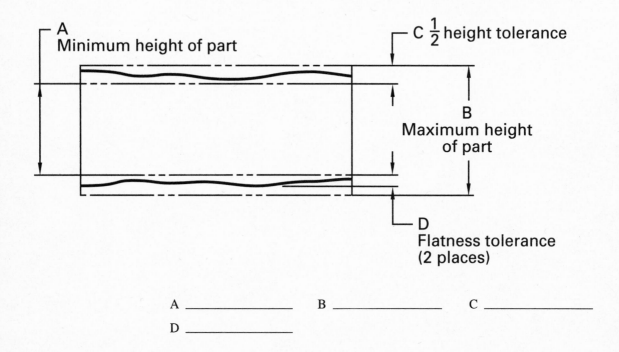

A _____   B _____   C _____

D _____

Chapter 9 Review Problems

Part I

Write the letter of the best answer in the space at left.

_____ 1. What does circularity control?

      (a) features   (b) surfaces   (c) elements   (d) datums

_____ 2. Does a circularity tolerance have to be within the dimensional limits of size?

      (a) it may   (b) it may not   (c) it must be   (d) it must not

_____ 3. A circularity tolerance always applies _____ .

      (a) FIM   (b) at LMC

      (c) RFS   (d) at MMC

_____ 4. What is the most likely shape of the tolerance zone for circularity?

      (a) a cylinder   (b) a radius

      (c) an ellipse   (d) a ring or washer

_____ 5. Circularity error is what type of distance between two concentric perfect circles?

      (a) radial   (b) diametral   (c) angular   (d) circular

_____ 6. A circularity tolerance can be what type of tolerance curled into a circle?

      (a) concentricity   (b) cylindricity

      (c) straightness   (d) flatness

_____ 7. For circularity to be most accurately inspected, what is a part placed on?

      (a) on a turntable   (b) on lathe centers

      (c) in a vee-block   (d) on a surface plate

Chapter 9 Review Problems

Part II

All the dimensions in these problems are in metric units (millimeters).

1. In the space below, draw the feature control frame for a circularity tolerance of .05 mm.

2. In the space below, draw two views of a cylinder 15 mm in diameter and 10 mm long. Space the views about 40 mm apart. Omit dimensions. Add a circularity tolerance of .05 mm.

3. In the drawing below, with no circularity control specified, what is the maximum permissible circularity error of the cylindrical surface?

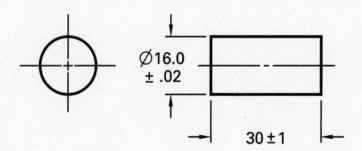

Ø16.0 ± .02

30 ± 1

4. In the space below, redraw the drawing for problem 3 and add a circularity tolerance of 0.01 mm to the cylindrical surface.

5. Shown below is one possible enlarged interpretation of the drawing for problem 4. Given the size and circularity tolerance of the correct drawing and this interpretation, fill in the values of dimensions A through D in the spaces provided.

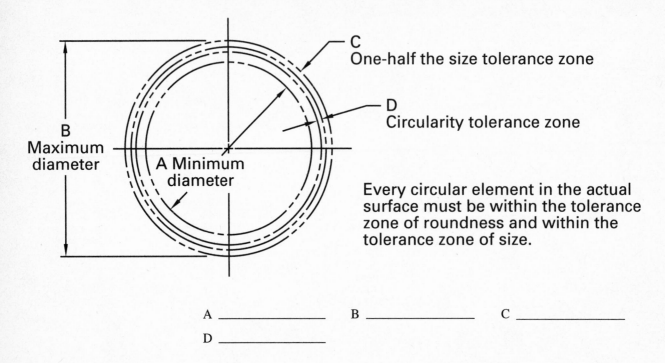

C
One-half the size tolerance zone

D
Circularity tolerance zone

Every circular element in the actual surface must be within the tolerance zone of roundness and within the tolerance zone of size.

B
Maximum
diameter

A Minimum
diameter

A _____   B _____   C _____

D _____

Chapter 10 Review Problems   Part I

Write the letter of the best answer in the space at left.

_____   **1.** Cylindricity is a combination of circularity and _____ .

                           (a) flatness   **(b)** straightness
                           (c) parallelism   **(d)** perpendicularity

_____   **2.** Circularity error is what type of distance between two concentric perfect circles?

                           (a) radial   **(b)** diametral   **(c)** total   **(d)** angular

_____   **3.** What is the most likely shape of the tolerance zone for cylindricity?

                           (a) doughnut   **(b)** globe   **(c)** tube   **(d)** cylinder

_____   **4.** A cylindrical tolerance _____ be within the dimensional size limits of the part

                           (a) may   **(b)** may not   **(c)** must   **(d)** must not

_____   **5.** A cylindricity tolerance always applies _____ .

                           (a) FIM   **(b)** at LMC
                           (c) RFS   **(d)** at MMC

_____   **6.** A cylindricity tolerance can be what type of tolerance curled into a circle?

                           (a) straightness   **(b)** flatness   **(c)** concentricity   **(d)** profile

_____   **7.** What must the longitudinal elements be checked for in cylindricity?

                           (a) concentricity   **(b)** straightness
                           (c) flatness   **(d)** coaxiality

Chapter 10 Review Problems   Part II

1. In the space below, draw the feature control frame for a cylindricity tolerance of .0005 inch.

2. In the space below, draw two views of a cylinder .63 inch in diameter and 1 inch long. Space the views about 1.5 inches apart. Omit dimensions. Add a cylindricity tolerance of .003 inch.

3. In the drawing below, with no cylindricity control specified, what is the maximum permissible cylindricity error?

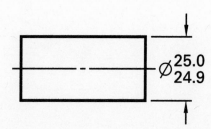
Ø 25.0
  24.9

4. In the space below, redraw the drawing for problem 3 and add a cylindrical tolerance of .01 inch.

Chapter 11 Review Problems   Part I

Write the letter of the best answer in the space at left.

_____   **1.** What does profile of a line control?

(a) chords   (b) radii   (c) elements   (d) arcs

_____   **2.** Unless otherwise specified, how is profile assumed to be applied?

(a) unilateral   (b) bilateral   (c) unidirectional   (d) aligned

_____   **3.** When a phantom line and a dimension are used to show a profile tolerance, how is it applied?

(a) unilaterally   (b) bilaterally   (c) unidirectionally   (d) aligned

_____   **4.** Profile of a surface tolerance controls the profile in _____ .

(a) one direction   (b) two directions
(c) three directions   (d) one curved line

_____   **5.** What must be specified on the drawing when a profile tolerance is related to another feature of the part?

(a) unilateral tolerance   (b) bilateral tolerance
(c) position tolerance   (d) datum

_____   **6.** What is the leader directed to when profile of a surface is used to control coplanarity?

(a) an extension line between the surfaces
(b) the larger surface
(c) a dimension line locating the surfaces
(d) the surface that is to be machined first

_____   **7.** What may profile of a surface control?

(a) form   (b) orientation
(c) position   (d) all of the above

Chapter 11 Review Problems

Part II

1. On the drawing below, show that the *curved elements* are within .1 mm total bilateral tolerance of the desired profile. The profile tolerance applies between points X and Y.

2. Redraw below the figure in problem 1 to make the tolerance only on the inside of the true profile.

4. On the drawing below, show that the entire curved surface is to have a bilateral profile tolerance of 0.1 mm.

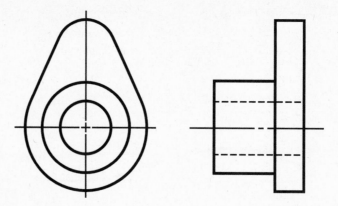

5. On the same cam, repeated below, show that the entire curved surface is to have a profile tolerance of .05 mm *outward* only of the desired profile.

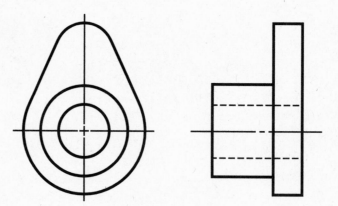

Chapter 12 Review Problems   Part I

Write the letter of the best answer in the space at left.

_____   1. Which of the following geometric characteristics does not require a
                  datum?
                  (a) cylindricity   (b) parallelism
                  (c) perpendicularity   (d) concentricity

_____   2. How many forms may parallelism take?
                  (a) two   (b) three
                  (c) four   (d) five

_____   3. Which symbol is placed in front of the geometric tolerance when the
                  tolerance zone has a cylindrical shape?
                  (a) parallel   (b) cylindricity   (c) diameter   (d) area

_____   4. How is a parallelism tolerance applied when no modifier is specified?
                  (a) at MMC   (b) at LMC
                  (c) RFS   (d) at virtual condition

_____   5. If a surface is controlled by a .002 parallelism tolerance, what is the
                  maximum flatness error?
                  (a) .0005   (b) .001   (c) .002   (d) .004

_____   6. When the MMC symbol is specified for a parallelism tolerance, the
                  controlled feature is _____ .
                  (a) parallel to a datum   (b) as large as possible
                  (c) as accurate as possible (d) a size featurel

Chapter 12 Review Problems   Part II

1. A surface of a part is parallel with the opposite surface within .005. In the space below, draw the appropriate feature control frame.

2. A hole in a part is parallel with the flat base of the part within .002, regardless of the produced size of the hole. In the space below, draw the appropriate feature control frame.

3. In the drawing below, the upper hole at MMC should be parallel with the lower hole within .002 diameter, regardless of the produced size of the lower hole. Add the appropriate symbols to the drawing.

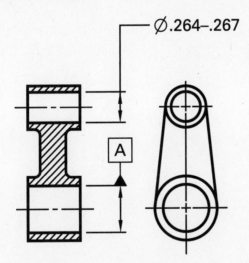

Ø.264–.267

4. In problem 3, the parallelism tolerance varies with the produced size of the upper hole. Complete the table below, showing this variation.

Feature Size	Diameter Tolerance Zone Allowed
.264	.002
.265	_____
.266	_____
.267	_____

Chapter 13 Review Problems

Part I

Write the letter of the best answer in the space at left.

_____ **1.** Which of the following is not a possible form for perpendicularity?

    (a) parallel lines   **(b)** parallel planes
    (c) parallel curves   **(d)** cylinder

_____ **2.** What must the form of the tolerance zone be before a diameter symbol is placed in front of the geometric tolerance?

    (a) cylinder   **(b)** circle   **(c)** arc   **(d)** axis

_____ **3.** If no modifier is specified in the feature control frame, how is perpendicularity applied?

    (a) at LMC   **(b)** at MMC
    (c) at virtual condition   **(d)** RFS

_____ **4.** What type of feature can have MMC applied with perpendicularity?

    (a) a tab   **(b)** a slot   **(c)** a keyway   **(d)** all of the above

_____ **5.** What is the allowance based on, for two mating parts, when they are controlled with perpendicularity?

    (a) MMC   **(b)** LMC   **(c)** RFS   **(d)** virtual condition

_____ **6.** What additional characteristic is controlled when perpendicularity is applied to a plane surface?

    (a) surface finish   **(b)** circularity
    (c) parallelism   **(d)** flatness

Chapter 13 Review Problems

Part II

1. A surface of a part is perpendicular to another surface within .003. In the space below, draw the appropriate feature control frame.

2. A hole in a part is perpendicular with the flat base within .002, regardless of the produced size of the hole. In the space below, draw the appropriate feature control frame.

3. In the part shown below, the hole at its MMC is perpendicular with the cylinder, RFS, within .004. Add an appropriate dimension and feature control frame.

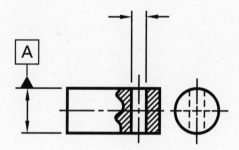

4. In problem 3, assume that the hole is dimensioned .250–.255 diameter. The perpendicularity tolerance will vary with the produced size of the hole. Complete the tolerance variation table below for every one-thousandth variation of the hole size.

Produced Size of Hole	Perpendicularity Tolerance
.250	⌀.002
_____	_____
_____	_____
_____	_____
_____	_____
_____	_____

5. In problem 4, what effect does the produced size of the cylinder have on the perpendicularity tolerance?

_____

6. In the drawing below, show that the boss is perpendicular to the base within .002 at MMC.

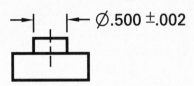

$\varnothing .500 \pm .002$

7. For the drawing in Problem 6, make up a tolerance variation table for five possible produced sizes of the boss. Use your own headings.

8. The two drawings below show a part with a tab and a part with a mating slot.

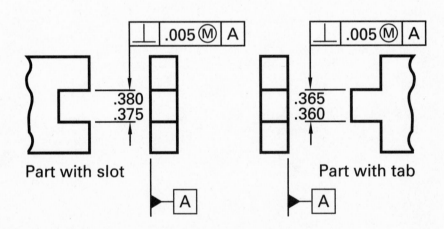

Part with slot   Part with tab

(a) What is the MMC of the slot? _____

(b) What is the MMC of the tab? _____

(c) What is the virtual size of the slot? _____

(d) What is the virtual size of the tab? _____

(e) What is the minimum clearance between the tab and the slot at the tightest fit? _____

Chapter 14 Review Problems

Part I

Write the letter of the best answer in the space at left.

_____ 1. What is the shape of the tolerance zone when the tolerance is
             expressed in degrees?
             (a) fan-shaped   (b) uniform   (c) radial   (d) conical

_____ 2. What is the shape of an angularity tolerance zone?
             (a) angular   (b) fan-shaped   (c) uniform   (d) radial

_____ 3. Which of the following is not a possible form for angularity?
             (a) parallel lines   (b) parallel planes
             (c) parallel curves   (d) cylinder

_____ 4. What is the connection between datums and angularity tolerances?
             (a) may use   (b) should use
             (c) should not use   (d) must use

_____ 5. How is angularity applied unless otherwise specified?
             (a) at LMC   (b) at MMC
             (c) RFS   (d) at virtual condition

_____ 6. Angularity of size features are more often controlled using which
             geometric control?
             (a) parallelism   (b) perpendicularity
             (c) position   (d) profile of a surface

1. In the part shown below, the slanted surface is held at 30° to the base within .005. Apply an angularity tolerance for this condition.

2. In the part shown below, the hole, regardless of feature size, is held at 60° to the base within .005. Apply an angularity tolerance for this condition.

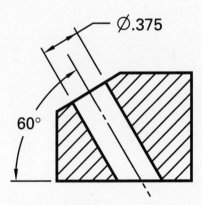

3. In the space below, draw the correct feature control frame for the drawing in problem 2 if the angularity tolerance were held at the MMC of the hole.

4. A .375–.380 diameter hole has an angularity tolerance of .006 at MMC relative to another hole, RFS. In the space below, draw a feature control frame to specify this situation.

5. When the .375–.380 diameter hole in problem 4 is *not* produced at its MMC, the angularity tolerance will not be .006. Make up a tolerance variation table below, giving the permissible angularity for each produced hole size, in thousandths.

Produced Size of Hole	Angularity Tolerance
_____	_____
_____	_____
_____	_____
_____	_____
_____	_____
_____	_____

6. In the drawing in problem 1, the flatness of the slanted surface is not specified. What is the maximum permissible out-of-flatness of this surface?

_____

Chapter 15 Review Problems

Part I

Write the letter of the best answer in the space at left.

_____ 1. What word is generally used to describe two circular features in line with each other?

  (a) concentricity   (b) alignment
  (c) coaxiality   (d) runout

_____ 2. In addition to concentricity and runout, which geometric characteristic can control the axial alignment of a hole?

  (a) location   (b) position
  (c) cylindricity   (d) circularity

_____ 3. Concentricity error is the amount by which the axes of two regular solids are _____ .

  (a) out of line   (b) in alignment   (c) curved   (d) irregular

_____ 4. Which of the three geometric characteristics used to control coaxiality is used the least?

  (a) position   (b) runout
  (c) circularity   (d) concentricity

_____ 5. The form of the concentricity tolerance zone is an imaginary cylinder about the axis of the _____ .

  (a) datum   (b) center   (c) feature   (d) outside diameter

_____ 6. How is concentricity always applied?

  (a) at LMC   (b) at MMC
  (c) RFS   (d) at virtual condition

_____ 7. How is the concentricity tolerance zone specified?

  (a) radius   (b) diameter   (c) total width   (d) radial width

Chapter 15 Review Problems   Part II

1. In the part below, the smaller diameter is to be concentric with the larger diameter within .002. Add a concentricity tolerance to the drawing.

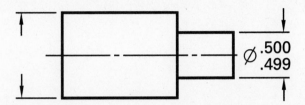

2. In the part below, the ∅.998–1.000 is to be concentric within .001 with the two small cylinders, used as one datum. Add a concentricity tolerance to the drawing.

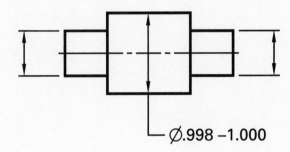

3. The drawing below is an incomplete interpretation of the concentricity control in the part shown in problem 2. Add an appropriate local note at each leader to complete the interpretation.

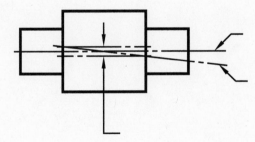

Chapter 16 Review Problems

Part I

Write the letter of the best answer in the space at left.

_____ **1.** How is runout inspected?

  (a) taking diametral measurements at 90°
  (b) rotating the part about an axis
  (c) using a micrometer
  (d) using a comparator

_____ **2.** What is the name for runout when it controls circularity, straightness, and coaxiality?

  (a) composite tolerance **(b)** complex tolerance
  (c) multiple tolerance   **(d)** position tolerance

_____ **3.** Circular runout is a control of circular _____ .

  (a) features   **(b)** axes   **(c)** surfaces   **(d)** elements

_____ **4.** Total runout is a control of all surface _____ .

  (a) elements   **(b)** features   **(c)** datums   **(d)** irregularities

_____ **5.** Circular runout is more commonly used because it is adequate for most designs and it is less expensive to _____ .

  (a) control   **(b)** measure   **(c)** specify   **(d)** analyze

_____ **6.** The form of the tolerance zone for circular runout is the radial distance between theoretical _____ .

  (a) cylinders   **(b)** elements   **(c)** circles   **(d)** radii

_____ **7.** The form of the tolerance zone for total runout is the radial distance between theoretical _____ .

  (a) cylinders   **(b)** elements   **(c)** circles   **(d)** radii

_____ **8.** What type of feature can be selected as a primary runout datum when the secondary datum is a relatively large cylindrical diameter?

  (a) axis   **(b)** flat face
  (c) pair of lathe centers   **(d)** element

_____ **9.** How is runout always applied?

  (a) at LMC   **(b)** at MMC
  (c) RFS   **(d)** at virtual condition

Chapter 16 Review Problems

Part II

1. The circular runout of a diameter is to be held to .002 relative to one datum axis established by two diameters, A and B. Draw the appropriate feature control frame.

2. The total runout of a diameter has to be maintained within .003 relative to another diameter and a perpendicular face. Draw an appropriate feature control frame. (Perpendicularity will not be specified.)

3. In the stepped shaft shown below, the circular runout of the conical surface relative to the smaller cylinder must be held within .001. Apply a circularity tolerance for this condition.

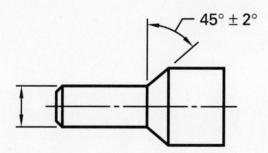

45° ± 2°

4. Under what two conditions is it best to use runout to control coaxiality rather than concentricity or position tolerance?

_____

_____

_____

Chapter 17 Review Problems   Part I

Write the letter of the best answer in the space at left.

_____ 1. What is the relationship between a feature axis or center plane relative to other features for position?

   (a) theoretically exact   (b) approximate
   (c) average   (d) accurate

_____ 2. What type of error does position allow?

   (a) permissible   (b) measurable
   (c) unilateral   (d) bilateral

_____ 3. What is the name for position when it controls location, parallelism, and perpendicularity at the same time?

   (a) composite tolerance   (b) complex tolerance
   (c) multiple tolerance   (d) profile tolerance

_____ 4. How is position usually applied when controlling circular features?

   (a) half-width   (b) total width   (c) radius   (d) diameter

_____ 5. When position is applied to rectangular features, what must remain within the tolerance zone?

   (a) center plane   (b) center line of symmetry
   (c) axis   (d) boundary

_____ 6. What would be an equivalent coordinate square tolerance zone for a .014 diameter position tolerance zone that would not allow interference between the mating parts?

   (a) .006   (b) .007   (c) .010   (d) .014

_____ 7. What type of dimensions are used with position?

   (a) chain   (b) basic   (c) limit   (d) toleranced

_____ 8. If a hole is positioned within a .014 diameter tolerance zone, how much can it be out-of-perpendicular?

   (a) .0035   (b) .007   (c) .014   (d) .021

1. A part contains a number of holes that are positioned relative to two edges of the part within ⌀.12 mm at MMC. Draw an appropriate feature control frame.

2. The drawing below is the interpretation for a part with a large hole held in position relative to two edges. Complete the local notes.

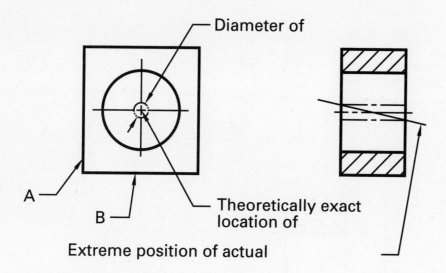

┌ Diameter of

A ⌐
B ⌐
└ Theoretically exact location of

Extreme position of actual

3. In the drawing below, the two holes are positioned relative to the left and bottom edges within ⌀.3 mm at MMC. Both datum surfaces are flat within .15 mm. Complete the drawing, adding symbolism as necessary and location dimensions for the holes.

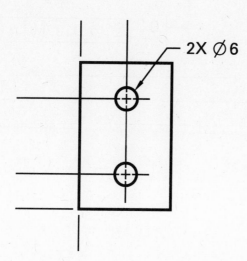

2X ⌀ 6

4. A strap has four 5 mm ±.05 mm punched holes in line, with a position tolerance of ∅.3 mm at MMC. Complete the tolerance variation table below for the possible produced sizes given.

Produced Size of Hole	Position Error Permissible
∅5.05 mm	_____
5.02	_____
5.00	_____
4.98	_____
4.97	_____

5. Relative to the example below, describe what happens to the four holes if the large hole is not at MMC.

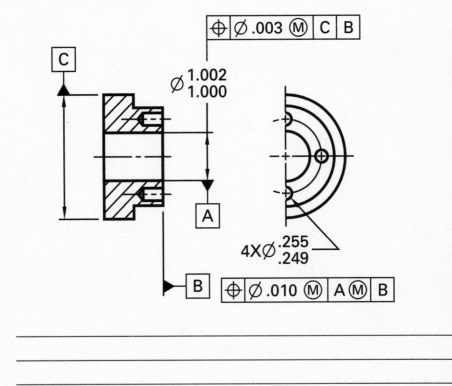

_____
_____
_____

The drawing of the spacer below is an application of the control of an edge distance by specifying LMC.

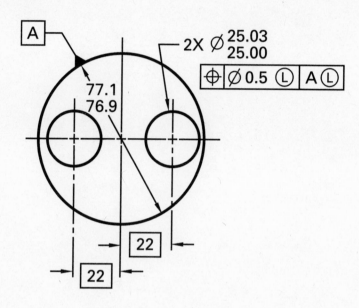

8. What is the edge distance when the holes and the datum are produced at their LMC sizes?

   _____

9. What is the maximum position tolerance for the two holes when they and the datum are *not* at LMC?

   _____

10. If the two holes and the datum were produced at their MMC sizes, what would be the edge distance?

    _____

Chapter 18 Review Problems

Write the letter of the best answer in the space at left.

_____ 1. What type of holes do both parts have in a floating fastener assembly?

(a) close-fitting   (b) press-fit   (c) clearance   (d) tapped

_____ 2. In a fixed fastener assembly of two parts, one of the mating parts has a clearance hole. What type of hole does the other part have?

(a) tight-fitting   (b) loose-fitting
(c) transition fit   (d) lubricated

_____ 3. Before a position tolerance can be specified, what must be calculated?

(a) clearance   (b) perpendicularity
(c) parallelism   (d) concentricity

_____ 4. What is the multiplication factor for the clearance in a floating fastener calculation?

(a) $\frac{1}{4}$ (b) $\frac{1}{2}$ (c) 1   (d) 2

_____ 5. What is the multiplication factor for the clearance in a fixed fastener calculation?

(a) $\frac{1}{4}$ (b) $\frac{1}{2}$ (c) 1   (d) 2

_____ 6. If a part is rejected because a size feature exceeds the MMC but actually fits the mating part, how can the position control be applied?

(a) projected tolerance zone
(b) zero position tolerance at MMC
(c) larger position tolerance
(d) smaller position tolerance

Chapter 18 Review Problems

1. Write a local note to replace the feature control frame in the drawing below.

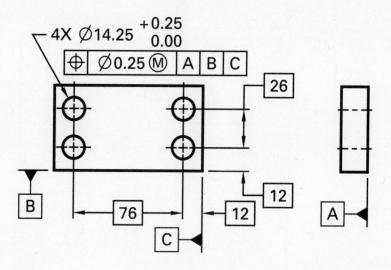

_____

_____

_____

2. To the drawing below, add a requirement that the axis of the hole be perpendicular to the top surface within ∅.3 mm at MMC and a requirement that the tolerance be projected 14 mm above the top surface.

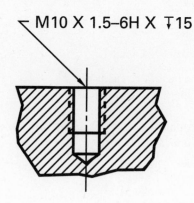

3. The next drawing is an application of zero position tolerance at MMC. The only variation allowed is in the size of the hole. Complete the tolerance variation table below for the given produced sizes of the hole.

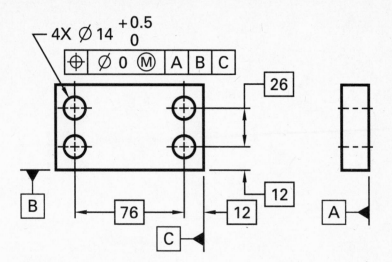

	Permissible
Produced Hole Size **(mm)**	**Position Tolerance** **(mm)**
∅14.0	_____
∅14.2	_____
∅14.5	_____

4. In the situation shown in problem 3, 13.9–14.0 mm screws are to be inserted in the four holes. What will be the allowance (tightest fit) between the screws and the holes when:

(a) the holes are produced at MMC? _____

(b) the holes are produced at ∅14.4 mm? _____

(c) What is the maximum possible clearance
(loosest fit) between the screws and holes? _____

5. The formula used to determine the clearance and position tolerance in the design of mating tabs and slots is similar to the floating fastener formula. Write the formula below, using

T_{tot} for total tolerance in both parts

W_s for width of slot at MMC

W_t for width of tab at MMC

_____

6. The drawing below shows a disk with two keyways located by position.

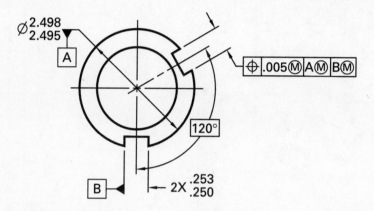

(a) What geometric characteristic (besides position) is actually controlled?

_____

(b) Write a local note that could replace the feature control frame.

_____

_____

_____

(c) Give one possible reason why the outside diameter is the more important of the two datums.

_____

7. For the drawing in problem 6, complete the tolerance variation table below for the possible produced sizes shown. (Hint: The total departure from MMC for all the sizes involved must be found.)

| Produced Size | | | Permissible |
Upper Slot	Lower Slot	Outside Diameter (O.D.)	Position Tolerance
.253	.253	2.498	_____
.252	.252	2.497	_____
.251	.251	2.496	_____
.250	.250	2.495	_____

8. The following questions refer to the drawing below of a plate with multiple patterns of features (two hole patterns).

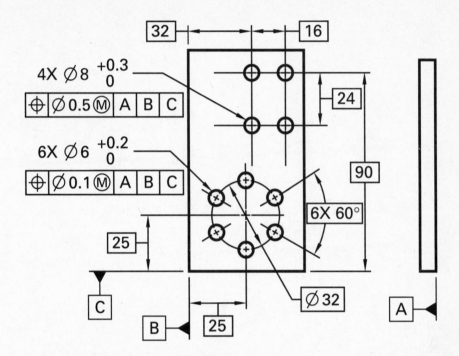

(a) Should the hole patterns be inspected as (check one) _____ separate patterns? _____ one composite pattern?

(b) Give the rule concerning multiple patterns of features that supports your answer to problem 8 (a).

_____

_____

_____

(c) How many gages will be used to inspect the location of the holes? _____

9. Write notes to interpret the composite feature control frame for the circular hole pattern in the drawing below.

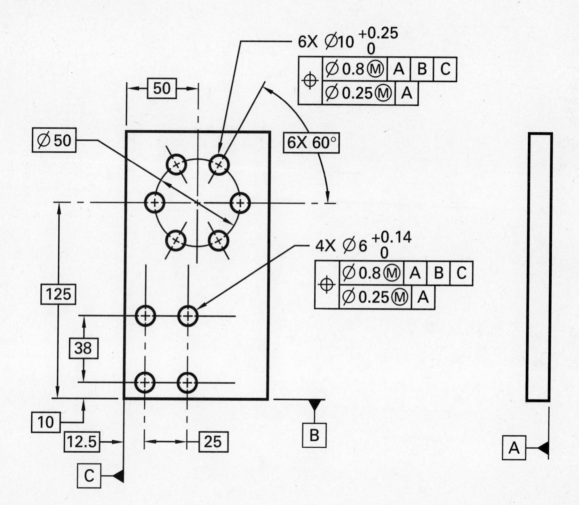

(a) The pattern location _____

_____

_____

_____

_____

_____

(b) The individual holes _____

_____

_____

_____

_____

_____

Chapter 19 Review Problems

Part I

Write the letter of the best answer in the space at left.

_____ 1. What do two or more coaxial features share?

            (a) center   (b) center plane   (c) axis   (d) datum

_____ 2. What is the shape of the tolerance zone used for coaxial position?

            (a) circle   (b) cylinder   (c) ring   (d) tube

_____ 3. How is a coaxial position tolerance specified?

            (a) radius   (b) diameter   (c) half-width   (d) total width

_____ 4. Which modifier is used for coaxiality to ensure all parts will assemble?

            (a) LMC   (b) MMC   (c) RFS   (d) virtual condition

_____ 5. In addition to controlling wall thickness, what else can an LMC modifier control?

            (a) edge distance   (b) basic dimension
            (c) location dimension   (d) assemblability

_____ 6. What is the total position tolerance equal to for a stepped shaft (two diameters) and stepped sleeve assembly?

            (a) sum of the maximum clearances
            (b) sum of the minimum clearances
            (c) difference between the larger diameters
            (d) difference between the smaller diameters

_____ 7. What is avoided when using zero position tolerancing?

            (a) unused position tolerance   (b) excess clearance
            (c) insufficient clearance   (d) excess runout

Chapter 19 Review Problems

Part II

1. In the seal body drawn below, the small diameter must be coaxial with the large diameter within ∅.001 when both diameters are at MMC. Add a position tolerance to complete the drawing.

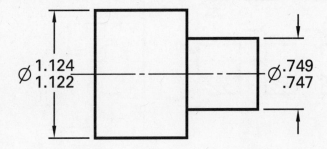

2. The outside diameter of a thin-wall sleeve must be positioned within ∅.1 mm at LMC relative to the inside diameter at LMC. Add a position tolerance to complete the drawing.

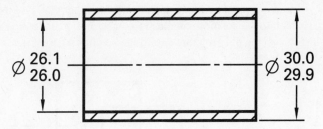

3. The small diameter of the balance block shown below must be coaxial within ∅.0005 at MMC with the large diameter, no matter what the produced size of the large diameter might be. Also, the large diameter at MMC must be positioned within ∅.010 relative to the height and width of the block at their MMC sizes. Add a position tolerance to complete the drawing.

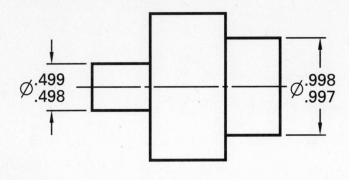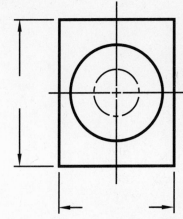

4. In the hinge body below, the four ∅10-mm holes are to be coaxial within ∅.1 mm at MMC and positioned within ∅.2 mm at MMC relative to two perpendicular flat surfaces. Add a composite feature control frame and datum feature symbols to express this design intent.

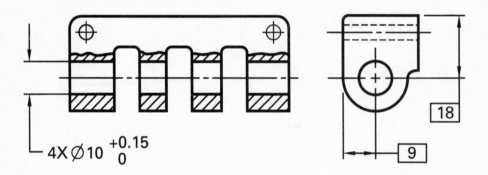

$4X \varnothing 10 \begin{smallmatrix} +0.15 \\ 0 \end{smallmatrix}$

5. Below are two drawings, one each of a socket and a mating plug. Use the three-step procedure to calculate the position tolerance for each part, making the tolerance for the socket larger than the plug tolerance. Add the position tolerances to the drawings.

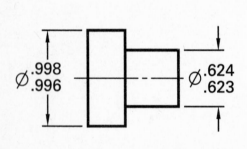

Plug

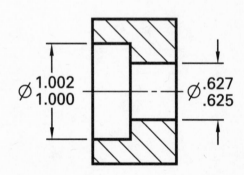

Socket

Step 1: Obtain the minimum clearances

Small diameters: _____ Large diameters: _____

Step 2: Add the minimum clearances to obtain the total position tolerance for both parts.

The sum is _____

Step 3: Divide the total position tolerance between the two parts.

Tolerance on socket: _____

Tolerance on plug: _____

Do not forget to add the position tolerances to the drawings.

6. The same seal body as in problem 1 is drawn below but with a zero position tolerance at MMC controlling coaxiality of the two diameters. Make up a tolerance variation table for two possible produced sizes: the MMC and the LMC for both diameters. Show the total departure (if any) from MMC resulting from the produced sizes. Show also the resulting permissible position error. Neatly letter appropriate headings.

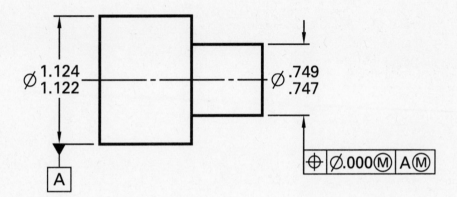

Chapter 20 Review Problems

Write the letter of the best answer in the space at left.

_____ 1. Symmetry is used in which of the following conditions?

    (a) mating parts   **(b)** interchangeable fits

    (c) force fits   **(d)** individual part

_____ 2. What is the main concern for the designer when using symmetry?

    (a) wall thickness   **(b)** strength

    (c) balance   **(d)** all of the above

_____ 3. Why is symmetry not commonly used?

    (a) it applies only to force fits   **(b)** it is expensive to inspect

    (c) it is hard to understand   **(d)** there are no tolerance calculations

Comprehensive Exercise 1

Definitions and Symbols

In the parentheses in the right-hand column, place the number of the matching term from the left-hand column. Use every term in the left-hand column.

1. Angularity
2. Basic dimension
3. Bilateral tolerance
4. Concentricity
5. Cylindricity
6. Datum
7. Flatness
8. Limits
9. Maximum material condition
10. Parallelism
11. Perpendicularity
12. Position
13. Profile of a line
14. Profile of surface
15. Regardless of feature size
16. Circularity
17. Runout, circular
18. Runout, total
19. Straightness
20. Tolerance
21. Unilateral tolerance
22. Symmetry

( ) Where the tolerance of form or position must be met, regardless of where the feature lies within its size tolerance.

( ) //

( ) ⌀⃫

( ) The theoretical value used to describe the exact size, shape, or location of a feature.

( ) ▱

( ) ∠

( ) One in which variation is permitted in both directions from the specified dimension.

( ) One in which variation is permitted in only one direction from the specified dimension.

( ) ⌒

( ) ⤦

( ) ◎

( ) ↗

( ) ⌖

( ) The condition of a part feature when it contains the maximum amount of material.

( ) ⊥

( ) —

( ) The total amount by which a dimension may vary.

( ) ○

( ) ⌒

( ) The maximum and minimum sizes indicated by a toleranced dimension.

( ) A surface indicated on the drawing from which measurements can be made.

( ) ≡

Name _____   Date _____

Comprehensive Exercise 2

Applications of Geometric Tolerances

Add feature control frames and datum identifying symbols to the drawing below to specify the following requirements. All symbols and feature control frames should be sized correctly.

Surface 1 to be circular within .005.

Surface 5 to be flat within .001.

Surface 2 to be cylindrical within .001 and positioned within .002 diameter relative to 5 and 1, regardless of feature size.

Surfaces 2, 3 and 4 to be positioned within .002 relative to surfaces 5 and 1 at MMC.

Surface 6 to be parallel within .002 to surface 5.

Runout of surface 7 within .003 relative to surfaces 5 and 1.

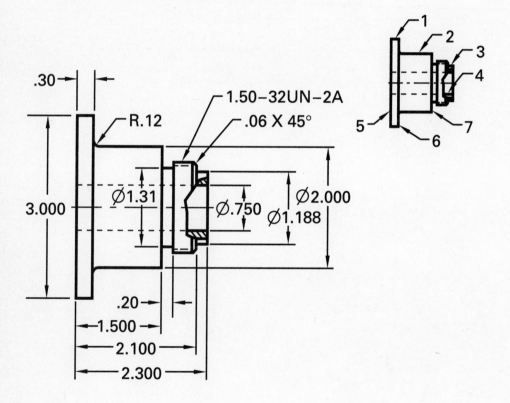

For each feature control frame and datum symbol on the previous page, write a note that might be used in place of the symbols. List them below by the numbers corresponding to the numbers shown on the drawing.

1. The cylindricity of the feature must be within .005. _____

2. _____

   _____

3. _____

   _____

4. _____

   _____

5. _____

   _____

6. _____

   _____

7. _____

   _____

Comprehensive Exercise 3

Calculations for Fasteners

1. A lid assembles onto a gear case with six $\frac{5}{16}$–18 cap screws. Using a position toler-ance of $\varnothing.015$ at MMC for the screw hole locations in both parts, calculate the size of the clearance holes in the lid. Selecting a drill size from the standard sizes listed at right. There are internal threads in the gear case. Perform all calculations neatly, starting with a formula.

Drill Sizes

.312
.316
.323
.328
.332
.339
.344
.348
.358

2. Three parts are held together with $\frac{1}{2}$–12 bolts. The clearance holes in all three parts are $\varnothing.531$. What position tolerance at MMC will be required in the drawings of the three parts? Perform all calculations neatly, starting with a formula.

Comprehensive Exercise 4

Calculations for Slots and Tabs

All the questions in this exercise pertain to the two drawings of mating parts below.

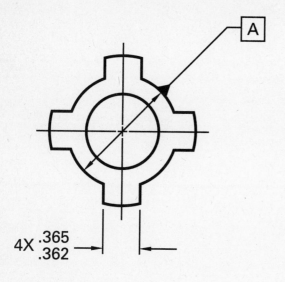

4X .365
.362

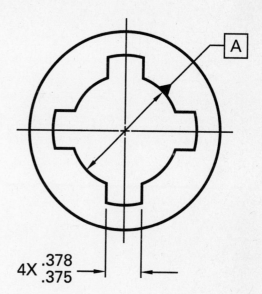

4X .378
.375

1. Calculate the position tolerance at MMC for the tabs and the slots, and add the appropriate feature control frames. Make the tab tolerance two-thirds of the slot tolerance. Show all calculations, starting with a formula.

   Tab tolerance: _____   Slot tolerance: _____

2. What is the maximum permissible position error when the tabs are at their LMC sizes? Show all calculations.

   _____

3. What would be the maximum permissible error when the tabs are at their LMC sizes if the position tolerance were specified RFS? Show all calculations.

   _____

Comprehensive Exercise 5

Calculations for Fit of Coaxial Parts

You have been asked to design the fit of the socket and mating plug shown in the two
drawings below.

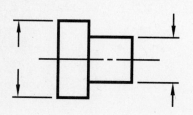

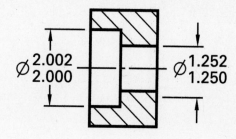

Plug–Scale: 1/2   Socket–Scale: 1/2

Given:

    The internal diameters of the socket:
        See drawing.
    The tightest fit with permissible out-of-coaxiality:
        Line-to-line
    The loosest fit with perfect coaxiality:
        .006
    The coaxiality tolerances:
        Plug   ⌀.001
        Socket   ⌀.002

Required:

    Decide whether the coaxiality tolerances and the datums should apply at MMC
    or RFS.
    Calculate limit dimensions for the plug diameters. (There is more than one correct
    answer.) Calculations can be done on scratch paper.
    Add the missing information to both drawings.
    You are not concerned with dimensions not related to the fit.

Comprehensive Exercise 6

Conversion—Coordinate Tolerances to Position Tolerances

The answers to all the problems in this exercise can be found by using the conversion tables in Appendix B. Round off to the nearest thousandth unless otherwise specified.

In the inspection of a baseplate, a coordinate measuring machine was used to measure deviation from basic hole location dimensions from two perpendicular datums (X and Y), and the data given below were recorded. Convert each pair of values to equivalent round position tolerance values.

Hole Number	Deviation X	Deviation Y	Position Tolerance
1.	.002	.004	_____
2.	.008	.006	_____
3.	.005	.001	_____
4.	.007	.003	_____

A large number of $\frac{5}{16}$-diameter holes are drilled in a housing. The print shows that the drill tolerance is $+.006 - .002$ and the holes must be positioned at MMC within .015 diameter. The deviations of the holes from their basic location dimensions are measured and recorded. The actual hole diameters are also measured and recorded. The table below presents the recorded data and provides spaces for other derived data to be filled in by the inspector. For each hole, obtain the equivalent position error at MMC, the *maximum* permissible position error, and indicate by a check mark whether the hole location is acceptable or not.

Hole Number	Deviation X	Deviation Y	Equivalent Position Error	Actual Hole Size	Maximum Permissible Position Error	Acceptable	Not Acceptable
5.	.006	.005	.016	.310	.015	_____	✓
6.	.007	.004	_____	.312	_____	_____	_____
7.	.002	.003	_____	.315	_____	_____	_____
8.	.008	.006	_____	.318	_____	_____	_____
9.	.007	.008	_____	.314	_____	_____	_____